International Pharmacy Conference (IPC 2017)

Product Authentication: Key Factor in Quality Control of Pharmaceutical Products

IOP Conference Series: Materials Science and Engineering
Volume 259

Yogyakarta, Indonesia
9 September 2017

ISBN: 978-1-5108-5308-9
ISSN: 1757-8981

Printed from e-media with permission by:

Curran Associates, Inc.
57 Morehouse Lane
Red Hook, NY 12571

proceedings.com

Some format issues inherent in the e-media version may also appear in this print version.

Printed by Curran Associates, Inc. (2017)

For permission requests, please contact the Institute of Physics
at the address below.

Institute of Physics
Dirac House, Temple Back
Bristol BS1 6BE UK

Phone: 44 1 17 929 7481
Fax: 44 1 17 920 0979

techtracking@iop.org

Additional copies of this publication are available from:

Curran Associates, Inc.
57 Morehouse Lane
Red Hook, NY 12571 USA
Phone: 845-758-0400
Fax: 845-758-2633
Email: curran@proceedings.com
Web: www.proceedings.com

Table of contents

Volume 259

International Pharmacy Conference UAD 2017 on Product Authentication: Key Factor in Quality Control of Pharmaceutical Products

9 September 2017, Yogyakarta, Indonesia

Accepted papers received: 18 October 2017
Published online: 27 November 2017

Preface

Papers

IOP Conf. Series: Materials Science and Engineering **259** (2017) 011001 doi:10.1088/1757-899X/259/1/011001

INTERNATIONAL PHARMACY
CONFERENCE 2017

PROCEEDING

The International Conference on
"Product Authentication : Key Factor in Quality Control
of Pharmaceutical Products"

September 9th, 2017
Sheraton Mustika Hotel, Yogyakarta, Indonesia

Held by :
Faculty of Pharmacy
Universitas Ahmad Dahlan Yogyakarta

IOP Conf. Series: Materials Science and Engineering **259** (2017) 011001 doi:10.1088/1757-899X/259/1/011001

PROCEEDING

INTERNATIONAL PHARMACY CONFERENCE (IPC) 2017

Theme :
"Product Authentication :
Key Factor in Quality Control of Pharmaceutical Products"

Keynote Speaker I
Representative of Ministry of Health Republic of Indonesia

Keynote Speaker II
Dr. Ir. H. Muslich, M.Si.
Head of Standard and Quality Assurance Div. LPPOM MUI

Speaker I
Prof. Dr. Rajendra SV., M.Pharm., Ph.D.
College of Pharmacy Krupanidhi Group of Institutions, India

Speaker II
Prof. Dr. Irwandi, Ph.D.
International Islamic University Malaysia (IIUM),
International Institute for Halal Research and Training (INHART), Malaysia

Speaker III
Prof. Dr. Winai Dahlan
Halal Center of Chulalongkorn University, Thailand

Speaker IV
Dr. Any Guntarti, M.Si., Apt.
Universitas Ahmad Dahlan, Indonesia

Saturday, September 9th, 2017
Sheraton Mustika Hotel, Yogyakarta, Indonesia

Organized by:
Faculty of Pharmacy, Universitas Ahmad Dahlan

IOP Conf. Series: Materials Science and Engineering **259** (2017) 011001 doi:10.1088/1757-899X/259/1/011001

Welcome Message from the Conference Chair

A product authentication is a process to ensure that a product is authentic or genuine. One of the criteria which becomes the object of authentication is halal products (food, cosmetics and medicine) through the existence of non-halal components of the products.

Faculty of Pharmacy, Universitas Ahmad Dahlan has successfully held an International Pharmacy Conference (IPC) with a theme "Product Authentication: Key Factor in Quality Control of Pharmaceutical Product" on 9th September 2017. This event was held as a forum to exchange information between researchers, academics, practitioners and policy holders mainly about halal authentication and regulations in national and international scope. There has been presented many kinds of research article as well.

We hope this proceeding can be a usefull knowledge for all readers. We would like to thank to all of the writers, speakers, presenters, participants and committees for the great contribution in this conference.

Best regards
Dr. Nanik Sulistyani, M.Si., Apt.
Conference Chair

IOP Conf. Series: Materials Science and Engineering **259** (2017) 011001 doi:10.1088/1757-899X/259/1/011001

Committee

Chairman :
Dr. Nanik Sulistyani, M.Si., Apt. *(Universitas Ahmad Dahlan, Indonesia)*

Secretary :
Citra Ariani Edityaningrum, M.Si., Apt. *(Universitas Ahmad Dahlan, Indonesia)*
Ginanjar Zukhruf Saputri, M.Sc., Apt. *(Universitas Ahmad Dahlan, Indonesia)*

Treasurer :
Ana Hidayati, M.Sc., Apt. *(Universitas Ahmad Dahlan, Indonesia)*
Haafizah Dania, M.Sc., Apt. *(Universitas Ahmad Dahlan, Indonesia)*

Scientific :
Iin Narwanti, M.Sc. *(Universitas Ahmad Dahlan, Indonesia)*
Dr. drh. Sapto Yuliani, MP. *(Universitas Ahmad Dahlan, Indonesia)*
Kintoko, M.Sc., Ph.D., Apt. *(Universitas Ahmad Dahlan, Indonesia)*

Program :
Dr. Laela Hayu Nurani, M.Si., Apt. *(Universitas Ahmad Dahlan, Indonesia)*
Hendy Ristiono, MPH., Apt. *(Universitas Ahmad Dahlan, Indonesia)*
Dr. Nining Sugihartini, M.Si., Apt. *(Universitas Ahmad Dahlan, Indonesia)*
Imaniar Noor Faridah M.Sc., Apt. *(Universitas Ahmad Dahlan, Indonesia)*
Hari Susanti, M.Si., Apt. *(Universitas Ahmad Dahlan, Indonesia)*
Dr. Wahyu Widyaningsih, M.Si., Apt. *(Universitas Ahmad Dahlan, Indonesia)*
Zainab, M.Si., Apt. *(Universitas Ahmad Dahlan, Indonesia)*
Nina Salamah, M.Sc., Apt. *(Universitas Ahmad Dahlan, Indonesia)*

Publication :
Lalu Muhammad Irham, M.Farm., Apt. *(Universitas Ahmad Dahlan, Indonesia)*
Azis Ikhsanuddin, M.Sc., Apt. *(Universitas Ahmad Dahlan, Indonesia)*

IOP Conf. Series: Materials Science and Engineering **259** (2017) 011001 doi:10.1088/1757-899X/259/1/011001

Editorial Boards

IOP Conf. Series: Materials Science and Engineering **259** (2017) 011002 doi:10.1088/1757-899X/259/1/011002

Peer review statement

All papers published in this volume of *IOP Conference Series: Materials Science and Engineering* have been peer reviewed through processes administered by the proceedings Editors. Reviews were conducted by expert referees to the professional and scientific standards expected of a proceedings journal published by IOP Publishing.

GC-MS analysis and antimicrobial activity determination of *Citrus medica* L. var proper leaf essential oil from South Sulawesi against skin pathogen microorganism

Aliyah, A Himawan, H Rante, Mufidah, D R Ningsih

Faculty of Pharmacy, Hasanuddin University, Makassar, Indonesia

E-mail: himawan@unhas.ac.id

Abstract. A research about *Citrus medica* L. var sarcodactylis had been conducted and it showed a significant antimicrobial activity, thus drive our curiosity to investigate the other variety from the same species, *Citrus medica* L. var proper. This research focuses in chemical compound study and antimicrobial activity screening against *Staphylococcus aureus*, *Propionibacterium acne*, and *Candida albicans* of *Citrus medica* L. var Proper leaves' essential oil. The essential oil is distillated from fresh leaves by hydrodestillation. The chemical compound was analysed using GC-MS instrument while the antimicrobial activity was tested using disk diffusion method. The results showed that the major component of the essential oil was Z-citral, citral and limonene compounds. The antimicrobial activity test results against the test microorganism are 9.15±0.15 mm, 11.15±1.3 mm and 8.02±0.48 mm, consecutively, for *Staphylococcus aureus*, *Propionibacterium acne* and *Candida albicans*.

Keywords: *Citrus medica* L. var Proper, antimicrobial, extract, essential oil

1. Introduction

Fruits from *Rutaceae* family (Citrus) are one of the word most produced fruits. Citrons are among the less known member of the family, which include oranges, mandarins, lemons and grapefruit, while they were the first fruits of the genus that reach Mediterranean region [1]. *Citrus medica* was a named given to C*itron* by *Linnaeus* possibly because its importance in pharmaceutical usage of the genus that time [2]. In general, *Citrus medica* L. had been used widely as traditional medicine and can be found in various traditional medicine literatures such as Ayurveda [3]. This plant belong to several variety like 'diamante', 'corsican', 'poncire commun', 'sarcodactylis' and 'rhobs-el-arsa' [1]. *Citrus media* L. var Proper is one of the variety that can be found in Indonesia and known for its empirical application in traditional medicine.

Investigation about antimicrobial activity of various Citrus plants had been conducted. A previous study shows that ethyl acetate extracts from various fresh and dried citrus peels were active against *Staphylococcus aureus* and *Escherichia coli* [4]. Another investigation focusing in antimicrobial activity of *Citrus medica* L. *var sarcodactylis* leaves' essential oil confirm its activity against *S. aureus, B subtillis, M. luteus and E. coli* with inhibition zone diameter 11±0.3 mm, 15±0.8 mm, 11±0.3 mm and 8±0.7 mm, consecutively [5]. Based on these findings, this research aimed to investigate further about antimicrobial activity form another variety of *Citrus medica* L., which is 'proper' variant as well as its physicochemical and phytochemical properties.

2. Materials and Methods

2.1. Plant material and authentication

Citrus medica L. var Proper leaves were collected from Makassar, South Sulawesi, Indonesia. The plant specimen was authenticated by Biology Department, Hasanuddin University.

2.2. Instrumentation

GC-MS-QP2010 Ultra Shimadzu is used in this study and the mass spectra database is used as comparison to identify the constituents.

2.3. Distillation

The fresh sample was subjected to modified hydrodistillation method with minimum water usage until the essential oil no longer appears in the condensate. The oil then dried with anhydrous sodium sulphate and stored in airtight container at 4°C [5].

2.4. GC-MS Analysis

The acquired essential oil was analyzed using GC-MS instrument with the standard protocol. Column condition: Rxi®-1 ms restech, 30x0.25 mm ID, 0.25 µm, column temperature 60°C, injection temperature 250°C, interface temperature 300°C. Helium was used as the carrier gas and dimethyl polysioxane is used as the stationary phase. One microliter of essential oil was diluted up to 2 mL with dichloromethane and the total runtime was 50 minutes. All the mass spectrum acquired was compared with the data base

2.5. Antimicrobial activity determination

Essential oil was tested against *Staphylococcus aureus*, *Propionibacterium acne* and *Candida albicans* using disk-diffusion method. The bacteria were cultivated on Nutrient Agar (Merck) while the yeast was grown in Potato Dextrose Agar (Merck) plate. Into a standard 6 mm blank disc (Oxoid), 20 µL samples were introduced with a micropipette and directly put onto the agar surface after sample introduction. All the plate then incubated in their respective temperature for 24 hours before inhibition zone diameter measurement. Standard antibiotic disks (Ampicillin and Nystatin) are used as positive control [6].

3. Results and Discussion

3.1. Chemical constituent of the essential oil

The GC-MS results showed that there are 17 peaks (Figure 1) that indicates that amount of compound presence in the essential oil with the major component of the essential oil was Z-citral and citral compound followed by dL-limonene and cyclohexane compounds. The detail was shown in the Table 1.

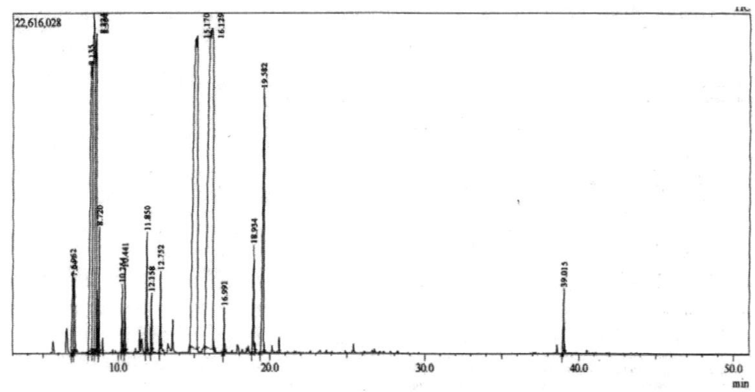

Figure 1. Chromatogram profile of the essential oil.

Table 1. Chemical components of the essential oil

No.	Name	RT	Area %
1	6-Methyl-5-hepten-2-one-β -myrcene	6.962	1.74
2	β-Myrcene	7.047	1.28
3	dL-Limonene	8.135	11.54
4	Cyclohexene, 1-methyl-4-(1-methylethenyl)	8.324	11.47
5	dL-Limonene	8.504	8.41
6	1,3,6-Octatriene,3,7 dimethyl-, (E)- (cas)	8.720	0.96
7	Nonanal (cas) n-nonanal	10.254	0.63
8	Linalool	10.441	0.84
9	Citronella	11.850	1.40
10	Phellandral	12.158	0.59
11	trans-Caran, 4,5 epoxy	12.752	0.87
12	Z-Citral	15.170	23.44
13	Citral	16.129	28.18
14	Undecanal (cas) hendecanal	16.991	0.45
15	Neryl acetate	18.934	1.47
16	Geranyl acetate	19.582	6.00
17	2-hexadecen-1-ol,3,7,11,15-tetramethyl-[R[R*R*-(E]]](cas) phytol	39.015	0.73

Table 2. Antimicrobial activity test results

Test microorganism	Inhibition zone (mm)	
	Citrus medica L. essential oil	Positive control
Staphylococcus aureus	9.15±0.15	15.27±2.43
Propionibacterium acne	11.15±1.33	13.29±3.21
Candida albicans	8.02±0.48	7.57±0.11

3.2. Discussion

Various parts of *Citrus medica* L. plants has been used traditionally to treat illness [7–9]. *Citrus medica* L. belong to Rutaceae family which is known, used and exploited for its essential oil content [10]. Essential oil (also known as volatile oil) from plants are known to possess antimicrobial activities [11].

The GC-MS analysis of the essential oil showed that the major components are citral, Z-citral and limonene compounds. A study reports a similar results about chemical components of *Citrus medica* L. leaves essential oil with limonene and citral as the major constituents [12]. The amount of limonene is somewhat low compared to the amount recovered from *Citrus medica* L. var Diamante with the same method [13]. The variation of volatile compound production of a plant can be affected by the geographical, environmental and physiological factors [14].

The antimicrobial activity maybe contributed by to the presence of limonene [5] and citral [15] in the essential oil. In the other hand, a study also reported a contradictory result about antifungal effect of citral containing essential oil [16]. There is a study which indicates that limonene and its enantiomers have antimicrobial properties against eight different test pathogens. The study reveals that stereochemistry play an important role in the effect of biological active essential oil which sometimes short-sighted. The same study also shows the possible antagonistic activity of chemical components of essential oil [17]. As a comparison, a research also reported that essential oil form *Citrus medica* L. var sarcodactylis possess antibacterial activity against *S. aureus* and *B. subtilis* with inhibition zone diameter 11±0.3 mm and 15±0.8 mm, respectively. The major components of the essential oil are limonene and citral (in the form of neral and geranial) [5].

A study that involve various citrus plants (orange, lemon and mandarin) shows that the antimicrobial activity of the essential oil may also be affected by the interaction of various chemical constituents of the oils (synergistic, addictive or antagonistic effects) [18]. Another research indicate that plant extract from various parts of the *Citrus medica* L. plant also have antimicrobial activity [19].

4. Conclusions

Essential oil from *Citrus medica* L. var proper fresh leaves has antimicrobial activity against *S. aureus*, *P. acne* and *C. albicans*. The major components of the essential oil are limonen and citral compounds.

Acknowledgement

The author would like to thank to Ministry of Research, Technology and Higher Education as the institution who grant this research through PUPT 2017 scheme.

References

[1] Tomi Â, Bessiere J, Casanova J, Lota M and Serra DDR 1999 Chemical composition of peel and leaf essential oils of *Citrus medica* L. and *C. limonimedica* Lush. *Flavour Fragr. J.* **166**(14) 161–6.

[2] Mabberley D J 2004 Citrus (Rutaceae): A review of recent advances in etymology, systematic and medical application. *BLUMEA* **49** 481–98.

[3] Panara K, Nishteswar K and Joshi K. A 2012 Review on phytochemical and pharmacological properties of citrus medica. *Int. J. Pharm. Biol. Arch.* **3**(6) 1292-97.

[4] Chanthaphon S, Chanthachum S and Hongpattarakere T 2008 Antimicrobial activities of essential oils and crude extracts from tropical *Citrus* spp. against food-related microorganisms. Songklanakairin *J. Sci. Technol.* **30** 125–31.

[5] Theanphong O, Songsak T and Mingvanish W 2008 Chemical composition and antimicrobial activity of the essestial oil from *Citrus medica* L . var . sarcodactylis (Sieber) Swingle Leaf. *Mahidol Univ. J. Pharm. Sci.* **35**(1–4) 57–61.

[6] Tavman A 2009 Antimicrobial activity of turkish citrus peel oils. *Pak. J. Bot.* **41**(6) 3207–12.

[7] Ahmed F 2017 Antiulcer activity of aqueous extract of *Citrus medica* Linn. fruit against ethanol-induced ulcer in rats *Advan. Biol. Res.* **6**(1) 24-29.

[8] Zhao P, Duan L, Guo L, Dou L L, Dong X, Zhou P, Li P and Liu E H 2015 Chemical and biological comparison of the fruit extracts of *Citrus wilsonii* Tanaka and *Citrus medica* L Article in Food Chemistry *Food Chem.* **173** 54-60.

[9] Sah AN, Joshi A, Juyal V and Kumar T 2011 Antidiabetic and hypolipidemic activity of *Citrus medica* Linn. seed extract in streptozotocin induced diabetic rats. *Pharmacogn J.* **3**(23) 80-84.

[10] Chutia M, Bhuyan PD, Pathak MG, Sarma T C and Boruah P 2009 Antifungal activity and chemical composition of *Citrus reticulata* Blanco essential oil against phytopathogens from North East India. *LWT-Food Sci. Technology* **42** 777–80.

[11] Solórzano-santos F and Miranda-novales MG 2015 Essential oils from aromatic herbs as antimicrobial agents. *Curr. Opin. Biotechnol.* **23**(2) 136–41.

[12] Bhuiyan M N I, Begum J, Sardar P K and Rahman M S, Cantonment POC 2009 Constituents of peel and leaf essential oils of *Citrus medica* L. *J. Sci. Res.* **1**(2) 387–92.

[13] Tundis R, Calabria U, Bonesi M, Calabria U and Calabria U 2011 Chemical composition and bioactivity of *Citrus medica* L. cv . Diamante essential oil obtained by hydrodistillation. *Nat. Prod. Res.* **25**(8) 789-99.

[14] Figueiredo A C, Barroso J G, Pedro L G and Scheffer J J C 2008 Factors affecting secondary metabolite production in plants: Volatile components and essential oils factors affecting secondary metabolite production in plants: volatile components and essential oils. *Flavour Fragr. J.* **23** 213–2261.

[15] Rojas-graü M A, Martin-belloso O, Mchugh T and Rojas-grau M A 2017 Effect of plant essential oil compounds on mechanical barrier and antimicrobial properties of alginate-apple puree edible films barrier and antimicrobial properties of alginate – apple puree edible films. *J. Food Eng.* **81**(3) 634-641.

[16] Nakahara K, Alzoreky N S and Yoshihashi T 2003 Chemical Composition and Antifungal Activity of Essential Oil from *Cymbopogon nardus* (Citronella Grass). *JARQ* **37** 249–52.

[17] Vuuren S F van and Viljoen A M 2007 Antimicrobial activity of limonene enantiomers and 1 , 8-cineole alone and in combination Antimicrobial activity of limonene enantiomers and 1 , 8-cineole alone and in combination *Flavour Fragr. J.* **22** 540–4.

[18] Espina L, Somolinos M, Lorán S, Conchello P, García D and Pagán R 2011 Chemical composition of commercial citrus fruit essential oils and evaluation of their antimicrobial activity acting alone or in combined processes *Food Control.* **22**(6) 896–902.

[19] Sah A N, Juyal V and Melkani A B 2011 Antimicrobial activity of six different parts of the plant. *Pharmacogn. J.* **3**(21) 80–3.

Relationship between emotional distress and quality of life on type 2 diabetes mellitus patients in Meranti island regency hospital

I N Faridah[1], D A Perwitasari[1], M Pusfita[2], H Jasman[1]

[1] Faculty of Pharmacy, Universitas Ahmad Dahlan, Yogyakarta, Indonesia

[2] Meranti Island Regency Hospital, Indonesia

Email: imaniar_apt@yahoo.com

Abstract. Type 2 Diabetes mellitus (T2DM) always got treatment for a long time so that it can affect the emotional distress and the quality of life. This study aimed to find the relationship between emotional distress with quality of life. This study used cross sectional design. DDS was used to measure patients emotional distress. EQ-5D was used to measure patients quality of life. Subjects of this study were T2DM patients ICD X.E11 with aged over 18 years old. A total of 80 patients participated in the study. The result showed that the mean of score on emotional burden was 2.985±0.678, physician distress was 2.650±0.801, regiment distress was 3.222±0.75 and interpersonal distress was 2.529±0.859. The result of the analysis showed that there was relationship between employment and interpersonal distress with index and correlation showed weak negative (r-0.212, -0.306) and significant relationship ($p<0.05$). The relationship between education and VAS showed weak positive correlation (r0.334) and significant relationship ($p<0.01$). The relationship between employment and emotional burden with VAS showed weak negative correlation (r-0.215, -0.251) and significant relationship ($p<0.01$). Employment and education was factors that affected the quality of life. High emotional distress can degraded the quality of life.

Keyword : Type 2 diabetes mellitus, emotional distress, quality of life

1. Introduction

Diabetes is a group of metabolic diseases characterized by hyperglycemia resulting from defected in insulin secretion, insulin action, or both. The chronic hyperglycemia of diabetes is associated with long-term damage, dysfunction, and failure of different organs [1]. This disease was caused by impaired glucose metabolism due to insulin deficiency both in absolute and relative [2]. T2DM is one of the most common types found of diabetes in more than 90-95% [1].

WHO predicted an increasing number of people with DM in Indonesia from 8.4 million in 2000 to about 21.3 million in 2030. This report showed an increasing number of people with DM 2-3 times by 2035. International Diabetes Federation (IDF) predicted an increasing number of people with DM in Indonesia from 9.1 million in 2014 to 14.1 million in 2035 [2].

Riskesdas report in 2013, Riau province was one of the area in Indonesia with prevalence of DM patients by 1.0% [1] and estimated 41,071 people [3]. Public health of Riau said that 2010 was the highest number of DM patients in the age group 45-54 years (191 cases), second age groups were 60-

69 (120 cases) and third age groups were 25-44 years (108 cases). Public health of Riau data in 2011 stated that in Pekanbaru the number of DM patients about 5.5% [4].

Individual's perception of the cognitive dimension can determine how and why patients can adhere to medical advice [5]. The cognitive representation consisted of five attributes: identity, time, consequence, cure control and cause. Cognitive representation was formed in parallel with the emotional representation. Emotional representations of feeling can be triggered by illness, for example, the diagnosis of chronic diseases can lead to anxiety or depression [6] . DM disease can have an impact on quality of life. The illness and the treatment had been undergone by a DM patient may affected the functional, psychological and social health and well-being of DM patients. It was can also affected the quality of life of DM patients [7] .

Based on the phenomenon above, researchers want to conduct further research about the relationship between the emotional distress with quality of life of patients with T2DM in Meranti island regency hospital.

2. Materials and Methods

The research was a cross sectional design. Patient data collection was conducted prospectively who was underwent outpatient treatment in Meranti island regency hospital in period February-April 2017. The sample size in this study was determined based on epiinfo. The sample size was calculated as followed:

$$n = [\mathrm{DEFF} * \mathrm{Np}(1 - \mathrm{p})]/ [(\mathrm{d2}/\mathrm{Z21} - \alpha/2 * (\mathrm{N} - 1) + \mathrm{p} * (-\mathrm{p})]$$

Population size (*N*) : 100
Hypothesized % (*p*) : 50%+/-5
Confidence limits as % of 100(absolute +/- %)(*d*) : 5%
Design effect (for cluster surveys-*DEFF*) : 1

Table 1. Number of sample based on confidence level

Confidence level %	Sampel size
95%	80
80%	63
90%	74
97%	83
99%	88
99.9%	92
99.99%	94

The 95% confidence level results in a sample of 80 people [8] .

The inclusion criterias in this study were patients with T2DM ICD No E11.00, E11.2, E11.3, E11.31, E11.32, E11.33, and aged over 18 years old. The exclusion criteria were T2DM patients underwent inpatient treatment at Meranti island regency hospital during the period of February-April 2017.

This study used cross sectional design. Diabetes Distress Scale (DDS) was used to measure patients emotional distress, and European Quality of life-5 Dimensions (EQ-5D) was used to measure patients quality of life.

The DDS consist 17 questions in 4 domains, emotional burden, physician distress, regiment distress, and interpersonal distress. Assessment was done by summing up the patient responses on the appropriate statements and divided it with the number of statements that exist in the domain [9].

IOP Conf. Series: Materials Science and Engineering **259** (2017) 012002 doi:10.1088/1757-899X/259/1/012002

Table 2. DDS questionnaire

No.	Domains	Question
1.	Emotional burden	7, 6, 9, 17, 8
2.	Physician distress	11, 12, 13, 4
3.	Regiment distress	14, 15, 10, 16, 5
4.	Interpersonal distress	3, 2, 1
Total		17

The EQ-5D questionnaire consist of 6 items covering 5 health dimensions: mobility, self-care, regular activity, pain/discomfort, and anxiety/depression, plus a global question to assess public health conditions [10]. Each dimension has 3 levels of assessment, which are no problem (value 1), some problems (value 2) and severe problem (value 3). The measurement result was the value of each dimension with a certain calculation was converted to EQ-5D index with the lowest value of -0.594 and the highest 1. The second part of EQ-5D contain Visual Analogue Scale (VAS) which to describe the subject perception about their quality of life by using the numeric scale 0 to 100. Which 0 is the lowest and 100 is the higest [11].

Univariate analysis used proportion or average value. The bivariate analysis used the linear regression, Mann-Whitney test for two-group data and Kruskal-Wallis test for three groups of data. Multivariate analysis used multiple linear regression test. This study received ethical approval from the Research Ethics Committee of Universitas Ahmad Dahlan.

3. Results and Discussion

Characteristics of patients with T2DM in Meranti island regency hospital can be seen in Table 3.

Table 3. Patient distribution

Category / characteristics	Total (n = 80)
Gender (%)	
Male	28 (35 %)
Women	52 (65 %)
Mean age (SD), year	57.41 (1.017)
Education (%)	
Lower education	54 (67.5%)
Higher education	26 (32.5%)
Employment (%)	
Unemployed	32 (40%)
Employed	48 (60%)
Income	
< Rp. 2.200.000	58 (72.5%)
> Rp. 2.200.000	22 (27.5%)
Marital status (%)	
Married	67 (83.8%)
Single/widow/divorced	13 (16.3%)
Family history (%)	
Yes	25 (31.3%)
No	55 (68.8%)
Type of treatment (%)	
Insulin	9 (11.3%)
Oral	65 (81.3%)
Insulin and oral combination	6 (7.5%)
Mean duration / duration of treatment of T2DM (SD), year	2.56 (0.194)
Companion disease (%)	
Microvascular	34 (42.5%)
Macrovascular	41 (51.3%)
Microvascular dan macrovascular	5 (6.3%)
Mean duration / duration of illness T2DM (SD), year	4.82 (0.576)

It showed that the incidence of T2DM occurred in women (65%) than in men (35%). It can be concluded that more women suffer from T2DM than men. This data was in line with previous research [12] in which the greater number of women was 53.3% and males 46.7%. Description of statistical emotional distress can be seen in Table 4. Based on the table, it can be seen that the mean ± SD result of DDS questionnaire measurement based on 4 domains. Based the mean result in 4 domains can be seen that the regiment distress was the largest which 3.222±0.75, and interpersonal distress was the smallest which 2.529±0.859. In the previous research reported that caregivers who was caring for a mentally impaired elder, who had been providing care for an extended time, and who had low social support was at high risk for psychological distress or depression [13].

Table 4. Statistical description emotional distress based on DDS

Domain	Mean	SD
Emotional burden	2.985	0.678
Physician distress	2.650	0.801
Regiment distress	3.222	0.750
Interpersonal distress	2.529	0.859

Analysis between characteristics with emotional distress patient in Table 5. Linear regression showed no significant relationship between characteristics with emotional distress (*p* value>0.05).

Analysis results between characteristics and quality of life in Table 6. Linear regression showed no significant relationship between characteristics with quality of life (*p* value>0.05).

Table 5. Analysis between characteristics and patients emotional distress

Characteristic	Emosional distress (DDS)							
	Emotional burden		Physician disstres		Regimen distress		Interpersonal disstres	
	x ± SD	*p* value	x ± SD	*p* value	x ± SD	*p* value	x ± SD	*p* value
Gender (n)								
Male (28)	3.05±0.648	0.553	2.830±0.778	0.087	3.264±0.824	0.367	2.582±0.827	0.621
Women (52)	2.95±0.697		2.55±0.803		3.200±0.715		2.500±0.883	
Education (n)								
Lower education (54)	3.092±0.729	0.015*	2.652±0.788	0.840	3.248±0.771	0.757	2.55±0.919	0.697
Higher education (26)	2.761±0.496		2.643±0.843		3.169±0.717		2.485±0.734	
Employment (n)								
Unemployed (32)	3.012±0.730	0.550	2.726±0.833	0.533	3.312±0.675	0.312	2.456±0.562	0.793
Employed (48)	2.966±0.647		2.598±0.783		3.162±0.797		2.577±1.013	
Income (n)								
< Rp. 2.200.000 (58)	3.017±0.705	0.684	2.599±0.766	0.241	3.162±0.761	0.159	2.575±0.919	0.564
> Rp. 2.200.000 (22)	2.900±0.607		2.783±0.890		3.381±0.711		2.404±0.678	
Family history (n)								
Yes (25)	2.928±0.624	0.584	2.550±0.930	0.196	3.056±0.751	0.066	2.516±0.927	0.698
No (55)	3.011±0.705		2.695±0.740		3.298±0.744		2.534±0.835	
Marital status (n)								
Married (67)	2.994±0.717	0.679	2.627±0.753	0.927	3.164±0.735	0.257	2.495±0.761	0.921
Single/widow/divorced (13)	2.938±0.435		2.769±1.043		3.523±0.785		2.700±1.279	
Type of treatment (n)								
Insulin (9)	2.733±0.547		2.444±1.333		3.067±0.959		2.422±1.337	
Oral (65)	3.003±0.621	0.328	2.665±0.742	0.291	3.264±0.730	0.404	2.533±0.806	0.543
Insulin and Oral combinations (6)	3.166±1.298		2.791±0.368		3.000±0.681		2.633±0.677	
Companion disease (n)								
Microvascular (34)	2.947±0.816		2.573±0.796		3.170±0.831		2.423±0.751	
Macrovascular (41)	3.053±0.567	0.330	2.640±0.786	0.290	3.209±0.709	0.310	2.539±0.924	0.198
Microvascular and macrovascular (5)	2.680±0.414		3.250±0.866		3.680±0.303		3.160±0.890	

**Mann Whitney* significant p *value* < 0.05

Table 6. Analysis between characteristics and patients quality of life

Characteristics	Quality of life (EQ-5D)			
	Index		VAS	
	x±SD	*p* value	x±SD	*p* value
Gender (n)				
Male (28)	0.767±0.261	0.714	81.07±16.852	0.520
Women (52)	0.751±0.287		76.92±22.011	
Education (n)				
Lower education (54)	0.720±0.298	0.059	72.78±21.405	0.000*
Higher education (26)	0.832±0.212		90.00±11.314	
Employment (n)				
Unemployed (32)	0.836±0.195	0.040*	83.75±18.272	0.030*
Employed (48)	0.704±0.311		74.79±21.036	
Income (n)				
< Rp. 2.200.000 (58)	0.739±0.293	0.464	75.34±21.620	0.030*
> Rp. 2.200.000 (22)	0.802±0.227		86.36±13.988	
Family history (n)				
Yes (25)	0.684±0.360	0.429	78.80±21.664	0.794
No (55)	0.789±0.225		78.18±19.916	
Marital status (n)				
Married (67)	0.789±0.235	0.092	79.70±19.304	0.295
Single/widow/divorced (13)	0.590±0.406		71.54±24.781	
Type of treatment (n)				
Insulin (9)	0.657±0.404	0.628	64.44±31.269	0.267
Oral (65)	0.766±0.254		80.77±17.793	
Insulin and Oral combinations (6)	0.799±0.319		73.33±21.602	
Companion disease (n)				
Microvascular (34)	0.834±0.203	0.222	78.24±21.387	0.504
Macrovascular (41)	0.700±0.308		79.51±19.615	
Microvascular and macrovascular (5)	0.696±0.365		70.00±21.213	

*Mann Whitney significant p value < 0.05
Index = part of the EQ-5D questionnaire
VAS = Visual analog scale, part of the EQ-5D questionnaire

The result of multiple linear regression can be seen in Table 7. In this study, variable employment affected the index, then the correlation relationship showed the weak negative (r -0.234) and $p<0.05$. Coefficient of determination showed the value of 0.045, means that employment can contribute the quality of life of 4.5%, the rest was explained by other factors. This means that the correlation of employment with quality of life was inversely proportional, meaning that employment can lead a decreasing score of quality of life. This data was in line with previous research [14] which employment affected the quality of life.

Table 7. Multiple linear regression analysis emotional distress with quality of life

Dependen t variable	Independent variable	Coefficient of determination (R^2)	Correlation relation (r)	*p* value
Index	Employment	0.045	-0.234	0.002*
	Interpersonal distress	0.094	-0.321	
VAS	Education	0.111	0.399	0.000**
	Employment	0.046	-0.217	
	Emotional burden	0.063	-0.321	

* multiple linear regression p value < 0.05
** multiple linear regression p value < 0.01
Index = part of the EQ-5D questionnaire
VAS = Visual analog scale, part of the EQ-5D questionnaire

IOP Conf. Series: Materials Science and Engineering **259** (2017) 012002 doi:10.1088/1757-899X/259/1/012002

Emotional distress may affect the index on the interpersonal distress. It showed from the negative correlation (r -0.321) and $p<0.05$. Coefficient determination showed 0.094, which interpersonal distrress contributed the quality of life of 9.4%, and the rest was explained by other factors. It can be concluded that the emotional correlation to quality of life was inversely proportional. The higher the emotional the worse the quality of life. In the previous research reported that a significant relationship between depression and the quality of life of respondents. The emergence of depression caused a significant decrease in the quality of life of DM patients [15].

Other variable, education can affect the VAS, with the correlation value (r 0.399) and $p <0.01$. Coefficient determination showed 0.111, which education contributed the quality of life by 11.1%, which means that the higher the education was better the quality of life. This was in line with the previous research [14], patients who had education above senior high school (lectures), the quality of life was better than the subject of research up to high school education. Education level was significantly related to quality of life [16]. The theory of behavioral science in the Health Educational Planning Diagnostic Approach states that quality of life was a way of assessing a person's health condition, in which health itself expressed physically, mentally and socially. Education was one of the predisposing aspects in which this predisposing was one that affect the behaviour of a person, so the level of education was one of a lot of factors that can be contributed to the quality of life of the patient. Another important factor was the non-behavioral factor in which it consists of things related to physiology, pharmacology, genetics and the patient's environment. Therefore, the result of research indicated that there was no correlation between level of education with quality of life was caused by very small association coefficient which can not significantly describe the relationship between them and it was possible there are other factors that have more dominant relation to quality of life [17].

In the employment variable, the value of the correlation of the employment indicated a weak negative correlation value (-0.217) $p<0.01$, it showed that employment can decrease the quality of life. The value of coefficient determination indicated the value of 0.046, meaning that the employment contributed to the quality of life of 4.6%, the rest was explained by other factors. Emotional distress can affect the VAS on the emotional burden, with the correlation value showed a weak negative correlation (r-0.321) and $p<0.001$. Coefficient determination showed 0.063, it showed that emotional burden can affect the quality of life by 6.3%. It can be concluded that the emotional distress has an inversely proportional correlation with quality of life. The higher the emotional distress, the worse the quality of life.

Based on previous research [18], family support was one of the most important element in DM patients, because the first and the most frequent interaction was the patient with the closest family. The presence of interpersonal support was very helpful for patients with T2DM to increase the confidence in their ability to do self-care [19]. Patients with good interpersonal support was had a sense of security and comfort so that will grow a sense of attention to themselves and increase the motivation to perform a disease management. This condition was aimed to prevent the emergence of stress in patients with T2DM. it can be understood if people with T2DM experience stress, and it will affect the body function. Stress was trigger to increase the cortisol in the body which will affect the increasing level of blood glucose by increasing gluconeogenesis, fatty and protein catabolism. In addition, cortisol was also interfere with the uptake of glucose by body cells that can affect blood glucose levels. This condition can cause an imbalance level of blood glucose and if this happens for a long time then the risk of complications will increase. In the end it can affect the quality of life of T2DM patients.

4. Conclusion

Employment and education was factors that affected the quality of life. High emotional distress can degraded the quality of life.

References

[1] ADA 2015 Classification and diagnosis of diabetes *Diabetes care* **38** 8-16.
[2] PERKENI 2015 *Konsensus Pengelolaan Dan Pencegahan Diabetes Melitus Tipe 2 di Indonesia* (Jakarta: Pb Perkeni).
[3] Kemenkes 2014 *Pusat Data dan Informasi* (Jakarta: Kementerian Kesehatan RI).
[4] DINKES 2010 *Profil Kesehatan Provinsi Riau* (Pekanbaru: Dinas Kesehatan Provinsi Riau)
[5] Jones CJ, Smith HE and Llewellyn CD 2016 A systematic review of the effectiveness of interventions using the common sense self-regulatory model to improve adherence behaviours *J. of Health Psychol.* **21** 2709-2724.
[6] Brandes K and Mullan B 2014 Can the common-sense model predict adherence in chronically ill patients? A meta-analysis *Health Psychology Review* **8** 129-153.
[7] Azila A A 2016 Gambaran kualitas hidup pasien diabetes mellitus tipe 2 di poli interna RSD dr. Soebandi Jember *Skripsi* Program Studi Ilmu Keperawatan Universitas Jember.
[8] Dea A G, Sullivan K M and Soe M M 2013 OpenEpi: source epidemiologic statistics for public health. accessed 30/10/2016 from www.OpenEpi.com.
[9] Fisher L, Hessler D M, Polonsky W H and Mullan J 2012 When is diabetes distress clinically meaningful? *Diabetes care* **35** 259-264 .
[10] Zeng Y C, Ching S S and Loke A Y 2010 Quality of life measurement in women with cervical cancer: implications for Chinese cervical cancer survivors *Health and quality of life outcomes* **8** 30.
[11] Harmaini F 2006 Uji keandalan dan kesahihan formulir European Quality of Life-5 Dimensions (EQ-5D) untuk mengukur kualitas hidup terkait kesehatan pada usia lanjut di RSUPNCM *Tesis* Fakultas Kedokteran Universitas Indonesia.
[12] Witasari U, Rahmawaty S and Zulaekah S 2009 Hubungan tingkat pengetahuan, asupan karbohidrat dan serat dengan pengendalian kadar glukosa darah pada penderita diabetes melitus tipe 2 *Jurnal Penelitian Sains & Teknologi* **10** 130-138.
[13] Baillie V, Norbeck J S and Barnes L E A 1988 Stress, social support, and psychological distress of family caregivers of the elderly *Nurs. Res.* **37** 217-222.
[14] Sari R M, Thobari J A and Andayani T M 2011 Evaluation quality of life of type 2 diabetic patients with oral antidiabetic therapy at ambulatory clinic in RSUP Dr. Sardjito *Journal of Management and Pharmacy Practice* **1** 35-42.
[15] Rantung J, Yetti K and Herawati T 2015 Hubungan self-care dengan kualitas hidup pasien diabetes mellitus (DM) di persatuan diabetes indonesia (Persadia) Cabang Cimahi *Jurnal Skolastik Keperawatan* **1.**
[16] Chiang C K, Peng Y S, Chiang S S, Yang C S, He Y H, Hung K Y, Wu K D, Wu M S, Fang C C and Tsai T J 2004 Health-related quality of life of hemodialysis patients in Taiwan: a multicenter study *Blood purification* **22** 490-498.
[17] Green L W, Kreuter M W, Deeds S G, Partridge K B and Bartlett E 2002 *Health Educational Planning: A diagnostic Approach* (California: Mayfield Publishing).
[18] Pratita N D 2012 Hubungan dukungan pasangan dan health locus of control dengan kepatuhan dalam menjalani proses pengobatan pada penderita diabetes mellitus tipe-2 *CALYPTRA* **1** 1-24.
[19] Antari G, Rasdini I and Triyani G 2012 Besar pengaruh dukungan sosial terhadap kualitas hidup pada penderita diabetes melitus tipe 2 di Poliklinik Interna RSUP Sanglah *Karya Tulis Ilmiah Strata satu.*

IOP Conf. Series: Materials Science and Engineering **259** (2017) 012003 doi:10.1088/1757-899X/259/1/012003

Actinomycetes of *Orthosipon stamineus* rhizosphere as producer of antibacterial compound against multidrug resistant bacteria

H Rante, R Yulianty, Usmar, N Djide, Subehan, R Burhamzah, M B Prasad

Laboratory of Microbiology Pharmacy, Faculty of Pharmacy, Universitas Hasanuddin, Makassar, Indonesia

E-mail: h_rante@yahoo.co.id

Abstract. The increasing case of antibiotic resistence has become an important problem to be faced in treating the infection diseases. The diversities of microbia, especially actinomycetes bacteria which originated from rizosphere soil of medicinal plant, has opened a chance for discovering the metabolites which can be used in solving the antibiotic resistant pathogenic bacteria problems. The aim of this research was to isolate the actinobacteria originated from medicinal plant rizosphere of *Orthosipon stamineus* as the producer of anti-multidrug resistances bacteria compounds. Three isolates of actinomycetes has been isolated from *Orthosipon stamineus* rhizosphere named KC3-1, KC3-2 and KC3-3. One isolate (KC3-3) showed big activity in inhibiting the test microbes by antagonistic test of actinomycetes isolates against *Staphylococcus aureus* and *Eschericia coli* antibiotic resistant bacteria. Furthermore, the KC3-3 isolate was fermented in Starch Nitrate Broth (SNB) medium for 14 days. The supernatant and the biomass of the fermentation yield were separated. The supernatant were extracted using ethyl acetate as the solvent and the biomass were extracted using methanol. The antibacterial activity test of ethyl acetate and methanol extract revealed that the extracts can inhibit the bacteria test up to 5% concentration. The ethyl acetate and methanol extracts can inhibit the bacteria test up to 5% concentration.

Keywords: Actinomycetes, *Orthosipon stamineus* rizosphere, antibacterial

1. Introduction

Study of discovering a new antimicrobial compound which is safe and potential is one of a challenge in pharmaceutical industry nowadays, especially in related to the increasing of oportunistic infection to the host [1]. The presence of antibiotic and other chemotherapeutic compounds becoming a relief thingeither nowadays or in the future, however it's still not enough yet. The main problem is there is a must for the presence of new medical agents which have a better potency in comparison to what have existed todays [2]. One of effort which can be done is doing a screening of bioactive compound towards any organisms or doing a modification towards the compounds which have existed todays.

Actinomycetes were a group of bacteria which is the most antibiotic-producing (70%) compared to fungi (20%) and bacteria (10%) [3], and have an important role in environment because it can doing biotransformatical and metabolism processes in a vast range. Beside, this bacteria is also the main microbia in micro-soil ecosystem [4] and it is the source of bioactive compound which is widely distributed in marine and terrestrial habitate [5]. Actinomycetes is abundant surrounds the high level plant rooting [6] and in rhizosphere soil approaching 40% in the total of soil microflora. Proteobacteria and actinomycetes were dominant population derived from any special of plant rhizosphere [7]. Microorganism abundant in rhizosphere because plant roots excrete exudate which contain organic compound such as amino acid, saccharine, and organic acid which were nutrients for the microorganism. Composition and the pattern of root exudat, plant species and soil type influencing the activity and the population of rhizosphere microbes.

2. Materials and Methods

2.1. Sampling
Rhizosphere soil samples were taken from medicinal plants in Makassar City of South Sulawesi Province. Samples were taken around root in 5-25 cm depth from ground level using sterile stainless still tools and then stored in tightly sealed containers which have been preserved with 70% alcohol, then taken to the laboratory for use in the study.

2.2. Actinomycetes isolation
The isolation of actinomycetes from soil samples was done by the method of spread and pour plate to obtain a single colony. Each dilution result was taken 1 mL and then spread into a Petri dish then SNA medium were poured then incubated at 30°C for 14 days. The colonies that grow and show actinomycetes are re-isolated to obtain a single colony. The pure colony was then inoculated into SNA media for use in subsequent tests

2.3. Testing of antibacterial activity with antagonistic test
Determination of antimicrobial activity was done by all isolates of actinomycetes grown into SNA media, then 7 days incubation actinomycetes on SNA media were placed on the surface of NA (Nutrient Agar) media which had been inoculated by test bacteria on separate plate for each test bacteria, then Incubated for 24 hours at 37°C. Each isolate was observed for its ability to inhibit the test bacteria characterized by the formation of clear zones around the isolates. A stable isolate forming a clear zone at 2 times test was chosen as the isolate for further testing

2.4. Fermentation of antibacterial active isolate
The pre-culture of the active isolates were made in a 500 mL erlenmeyer flask containing 100 mL of SNA liquid medium and incubated at 28°C for 3 days. Pre-culture (starter) is transferred into a 500 mL Erlenmeyer containing 100 mL of the same medium. Fermentation was carried out at a temperature of 28°C for 14 days under the conditions of the agitated at a rate of 150 rpm.

2.5. Extraction of metabolite compound
After fermentation for 14 days, microbial growth medium is filtered to separate biomass and fermentation liquids. The fermentation fluid is extracted 2 times with ethyl acetate (1: 1 v/v) solvent in a separating funnel for 20 min while the biomass is extracted with methanol. The extract obtained was evaporated and then stored in the desiccator for use in the next test.

2.6. Antibacterial activity test
One culture of rejuvenated bacteria on NA (Nutrient Agar) media was suspended into a tube containing 5 mL of NB (Nutrient Broth) medium and incubated for 24 hours at 37 C. The suspension of the bacteria was diluted using 0.9% sterile NaCl to its turbidity equivalent to the standard 0.5 McFarland solution (bacterial population $1x10^7$ CFU/mL - $1x10^8$ CFU/mL). The composition of McFarland's standard solution is $BaCl_2$ 0.048 M 0.5 mL and H_2SO_4 0.18 M 99.5 mL.

Antibacterial activity was determined by bioassay method based on [1] and [8] against test bacteria. The extract obtained was tested for the test bacteria in the following way: extract of ethyl acetate with certain concentration dissolved in organic solvent. A total of 20 μL maserate which has been known for its concentrate is inserted into the disc paper (6 mm in diameter). After all solvents have evaporated further disc paper is placed on the surface of the media that has been inoculated with test bacteria with streak method. All plates were incubated for 24 hours at 37°C, observed the presence of antibacterial activity characterized by a zone of growth inhibition of bacteria around the disc paper.

3. Results and Discussion

3.1. Isolation and selection of Orthosipon stamineus rhizosphere actinomycetes

Isolation of actinomycetes from rhizosphere *Orthosipon stamineus* was performed using the spread plate method. Using general media for actinomycetes isolation SNA (Strach Nitrate Agar). Starch Nitrate, starch casein, and Glycerolglycine are the most suitable medium for isolating actinomycetes bacteria, but Starch Nitrate is the most commonly used medium because it can isolate different types of actinomycetes based on different pigmentation and morphological characters so that the Starch medium Nitrate is the most suitable culture medium [9].

In order to distinguish between actinomycetes isolates and bacterial and fungal contaminants, we observed the morphological characteristics of actinomycetes, actinomycetes colonies showed a consistency of powders and adherence strongly to the medium and had slow growth, in contrast to other bacterial colonies that had slimy surfaces and growth fast. Actinomycetes isolates grown on SNA media are characterized by colonies that resemble fungal colonies but are more compact.

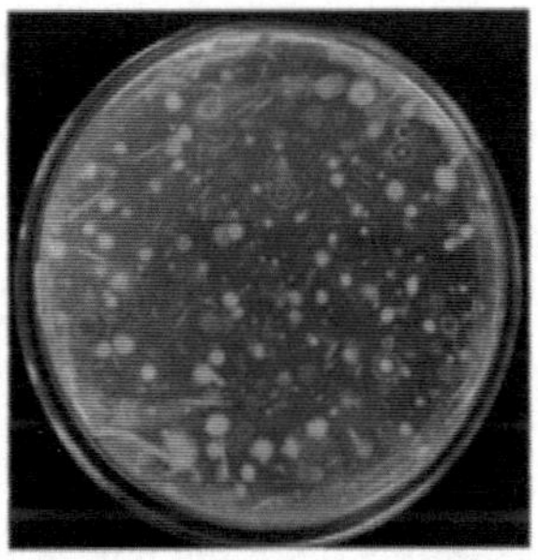

Figure 1. Microbia colony (A, B, C: Actinomycetes colony; a: Bacteria colony; b: Fungi colony)

The three isolates of actinomycetes obtained were separated and purified by streak on SNA medium then named KC3-1, KC3-2 and KC3-3 (Figure 2).

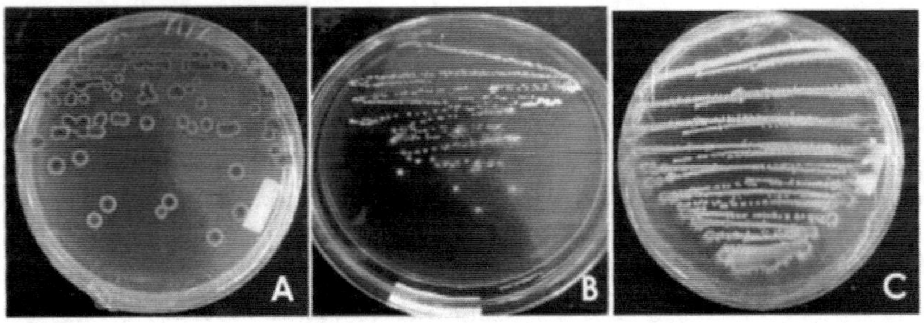

Figure 2. Actinomycetes isolate A : Actinomycetes isolate KC3-1; B. Actinomycetes isolate KC3-2; C. Actinomycetes isolate KC3-3

3.2. Antagonistic test of actinomycetes isolate

Antagonistic test is a qualitative test that is aimed to know the antibacterial activity of the pure *actinomycetes* isolates. The method used in antagonistic test was *agar block* method. Of the three isolates tested, only two isolates showed an inhibitor of *S. aureus* and *E. coli* bacteria, namely KC3-1 and KC3-3 isolates characterized by clear zones on agar that already contained test microbes. And isolate KC3-3 indicating the largest antibacterial activity so that this isolate is continued into the fermentation stage.

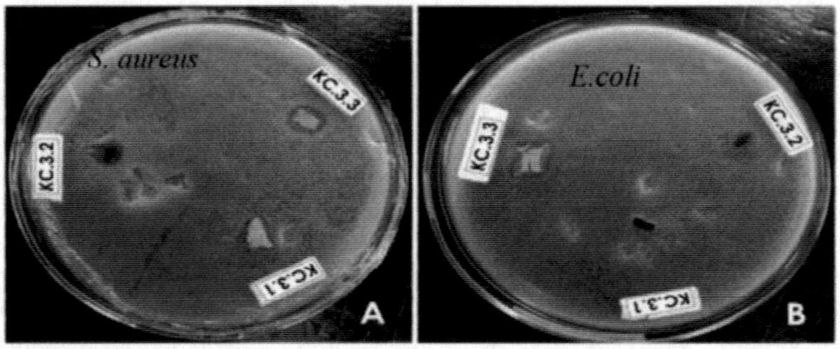

Figure 3. Antagonist test of isolate actinomycetes with zones of inhibition

3.3. *Fermentation and antibacterial activity test*

Fermentation was done to obtain the secondary metabolite from actinomycetes isolate. Fermentation was conducted on pure isolate which was 7 days old. Fermentation on KC3-3 isolate was conducted using SNB (*Starch Nitrate Broth)* medium and then incubated for 14 days. According to [10], On the 14th days of incubation, generally actinomycetes has entered the stationary phase. This is the phase of the bacteria do not show significant growth or the microorganism growth stop and the balance is present between the splitting cells and the dying cells. This is related to the nutrition which is reduced so there was a competition between microbes so that in this phase the microbes were producing antibacterial compound (secondary metabolite) to prevent the competition of nutrition and space. The fermentation yield (supernatant) were extracted using ethyl acetate twice (1:1 v/v), and biomass extracted using methanol. The extract obtained were tested for its antibacterial activity.

Antibacterial activity test result revealed that the supernatant and the biomass of fermentation extract of KC3-3 actinomycetes can inhibit the growth of test bacteria. The production of secondary metabolite of KC3-3 actinomycetes isolate was produce extracellularly to the fermentation media or supernatant, although in intracellular of the biomass was also present the secondary metabolite. This is based on antibacterial activity which was bigger than its supernatant liquid which has been extracted by organic solvent of ethyl acetate compared to the metabolite that is produced in biomass of cell (Figure 4). In the process of extracting fermentation or supernatant liquids is used ethyl acetate solvent, in addition to the greatest antibacterial activity in ethyl acetate extract also based on the research of [11] among the four types of solvents used hexane, chloroform, ethers and ethyl acetate, the most promising ethyl acetate extract as an antimicrobial with a large range of bacteria and fungi. [12] showed that metabolite extraction of actinomycetes using ethyl acetate was based on solubility and maximized antimicrobial activity. Ethyl acetate extract and methanol extract from fermentation of isolate actinomycetes KC3-3 were able to inhibit *S. aureus* and *E. coli* bacteria up to 5% (1 mg / paper disk). These results indicate that actinomycetes isolate of KC3-3 isolated from rhizosphere *Orthosipon stamineus* can be a potential antibiotic compounds producer.

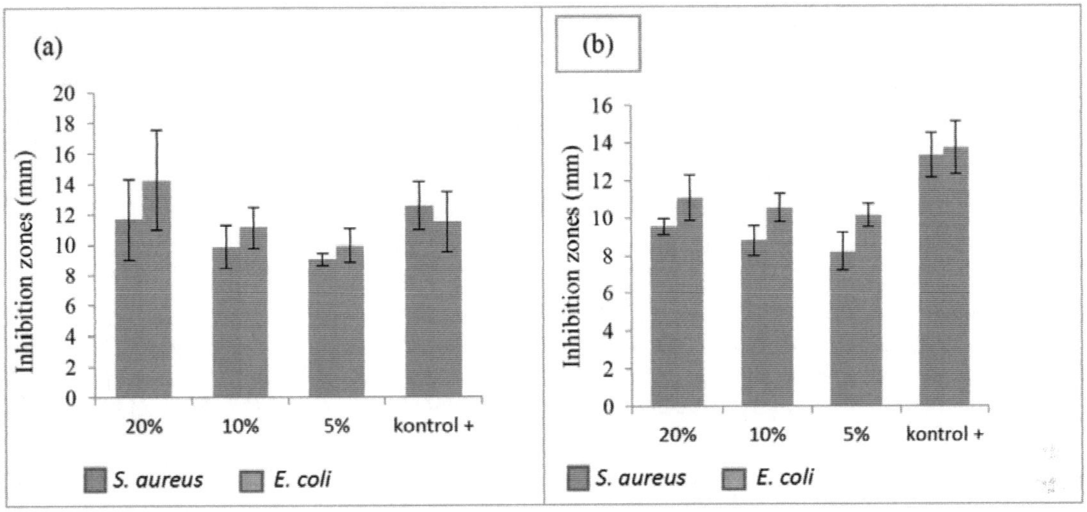

Figure 4. Result of antibacterial test of isolate actinomycetes KC3-3, (a) Ethyl acetate extract from supernatant, (b) Methanol extract from biomass.

4. Conclusions

Three bacteria isolates which isolated from *Orthosipon stamineus* rhizosphere were in group of actinomycetes named KC3-1, KC3-2 dan KC3-3. Actinomycetes of KC3-3 isolate was able to produce secondary metabolite which was active against *S. aureus* and *E. coli* up to 5 % concentration which were equivalent to 1 mg/paperdisk.

Acknowledgement

The authors thank to Ministry of Research, Technology and Higher Education of The Republic of Indonesia which has funding this project in PUPT 2017 scheme.

References

[1] Badji B, Zitouni A, Mathieu F, Lebrihi A and Sabaou N 2006 Antimicrobial compounds produced by Actinomadura sp AC104 isolated from an algerian sahara Soil *Can. J. Microbiol.* **55(4)** 373-382.

[2] Allen N E and Nicas T I 2003 Mechanism of action of oritavancin and related glycopeptides antibiotics *FEMS Microbiology Review* **26** 511-532.

[3] Atlas R 1997 *Principle of Microbiology* (USA: Wmc. Brown Publisher).

[4] Thangapandian V, Ponmurugan P and Ponmuruga N K 2007Actinomycetes diversity in the rhizosphere soils of different medicinal plants in Kolly Hills-Tamil Nadu, India, for secondary metabolite production *Asian J. Plant. Sci.* **6** 66-70.

[5] Pathom-Aree W, Stach J E M, Ward A C, Horikoshi K, Bull A T and Goodfellow M 2006 Diversity of actinomycetes isolated from challenger deep sediment (10,898 m) from the Mariana Trench *Extremophiles* 2006a **10** 181–9.

[6] Oskay M, Tamer A U and Azeri C 2004 Antibacterial activity of some actinomycetes isolated from farming soils of Turkey *Afric. J. of Biotech.* **3**(9) 441-446.

[7] Singh L S, Buruah I and Bora T C 2006 Actinomycetes of Loktat habitat: Isolation & screening for antimicrobial activity *Biotechnology* **5**(2) 217-221.

[8] Pandey B, Ghimire P and Agrawal V P 2004 Studies on the antibacterial activity of actinomycetes isolated from the Khumbu Region of Mt. Everest. International Conference on the Great Himala-yas: Climate, Health, Ecology, Management and Conservation, Kathmandu. Kathmandu University and the Aquatic Ecosystem Health and Management Society, Canada. January 12-15, 2004.

[9] Ghanem N B, Sabry S A, El-Sherif Z M, Abu El-Ela G A 2000 Isolation and enumeration of marine actinomycetes from seawater and sediments in Alexandria *J. Gen. Appl. Microbiol.* **46** 105–111.

[10] Mulyadi and Nanik S 2013 Aktivitas cairan kultur 12 Isolat actinomycetes terhadap bakteri resisten *KESMAS* **7**(2) 55-112.

[11] Arasu M V, Duraip&iyan V, Agastian P, Ignacimuthu S 2009 In vitro antimicrobial activity of *Streptomyces* spp. Eri-3 isolated from Western Ghats Rock Soil (India) *J. de Mycologie Medicale* **19** 22-28.

[12] Alim Al-Bari M A, Sayeed M A, Rahman M S, Mossadik M A 2006 Characterization and antimicrobial activities of a phthalic acid derivative produced by *Streptomyces bangladeshiensis*: A novel species collected in Bangladesh *Research J. Med. Medical. Sci.* **1**(2) 77-81.

Relationship between perception with the quality of life of T2DM patients in Dok II Jayapura Hospital

D A Perwitasari[1], I N Faridah[1], Y Kulle[2], M Yulistika[1]

[1]Faculty of Pharmacy, Universitas Ahmad Dahlan, Yogyakarta, Indonesia

[2]Dok II Jayapura Hospital, Indonesia

E-mail: diahperwitasari2003@yahoo.com

Abstract. Diabetes mellitus disease is one of the chronic diseases which can cause a fatal risk if its management is not appropriate. The purpose of this study was to determine the relationship between perception and quality of life in Dok II Jayapura hospital. This study used an observational research design with cross sectional approach with prospective sampling in diabetic patients in internal disease policemen who were underwent outpatient treatment at Dok II Jayapura hospital. Research instruments used EQ-5D (European Quality-5 Dimension) and B-IPQ (Brief Illness Perception Quetionare). Data analysis used was univariate analysis by using percentage or mean value, bivariate using T-test or Mann-Whitney test, and multivariate using multiple linear regression. There were 80 T2DM patients who met the inclusion criteria. Based on patient demographic data, there were 29 people (36.6%) male patients and 51 people (63.8%) female patients, with mean age of patient (55.79±10.52) year. Perception has correlation with quality of life influenced by index value on treatment control (6.73±1.475) and emotional response (3.11±2.199) and by visual analog scale on understanding (5.99±1.587), duration (6.50±2.968), and personal controls (6.20±1.641). Based on the results obtained to improve the quality of life of T2DM patients that is on the index value on the control of treatment and emotional response and on visual analog scale on the understanding, duration and personal control should be changed so that the quality of life of patients increases. The family history, social status and type of treatment factors also affect the quality of life.

Keywords: diabetes mellitus, perception, quality of life

1. Introduction

Diabetes mellitus (DM) is one of the most chronic illnesses in the world. According to research conducted by *the World Health Organization* [1], is estimated to reach 171 million people in 2000 and this figure will rise to 366 million in 2030. Meanwhile, Indonesia is a country that was ranked fourth, with total patient about 8.4 million people in 2000 and will increase to 21.3 million people by 2030. The highest prevalence of Diabetes Mellitus is found in Riau Province (1.0%), West Kalimantan (0.8%) and Papua (0.8%) [2].

B-IPQ (Brief-illness perception questionare) is a questionnaire designed to facilitate assesment used by patients to describe the patient's perception of the disease [3]. Therefore we expect that private control is a higher confidence can be measured by *the Brief-Illness Perception Questionare* and also can be associated with a better metabolic control.

IOP Conf. Series: Materials Science and Engineering **259** (2017) 012004 doi:10.1088/1757-899X/259/1/012004

According to previous research, which encourages the necessity of measuring the quality of life, especially in people with diabetes mellitus is because the quality of life is one of the main goals of care, if blood sugar levels can be controlled properly, then the physical complaints due to acute and chronic complications can be minimized or prevented. In addition, low quality of life and psychological problems can make heavier metabolic disorders, either directly through hormonal stress reactions, or indirectly through complications [4]. Knowledge of the patient's condition, the condition of the patient's illness, and family support can affect patient compliance behavior and then can affect the output of clinic patients [5]. Family knowledge and support are two main factors that great affect the adherence of diabetes mellitus patients in management [6].

2. Materials and Methods

The research was conducted prospectively at Dok II Jayapura hospital. This study was use a study design *cross-sectional,* prospective data collection on patients in Dok II Jayapura hospital. Population diabetes mellitus patients who do outpatient in polyclinic of disease in Dok II Jayapura hospital. The sample is diabetes mellitus patient who participated in the study, which corresponds to the inclusion criteria listed below man and woman patients included in the category of adults aged $\geq$ 18 years exclusion criteria diabetic patients underwent outpatient treatment at Dok II Jayapura hospital. Diabetes mellitus sufferers without and with complications. Sample is done by doing a screening through the laboratory. The selected sample is diabetes mellitus patient out patient's address recorded and contacted, then seen suitability with inclusion criteria.

In this study the research used the formula for calculating the large sample of diabetes mellitus patients in Dok II Jayapura hospital the sample size N=Population size of patient DM = 100 P = Hypothesis of % frequency of result factors in population if not stated otherwise = 50% D = Trust limit as % of 100 (absolute +/-%) = 5% DEFF = Design effect (for survey-Deff Cluster) =1 With the results of the level of 95%, the number of patients with DM obtained 80 samples of the population of respondents with diabetes mellitus in Dok II Jayapura hospital is 100. Data perceptions, behavior and *self-management* support was used a structured questionnaire that has been validated. The perception questionnaire will be filled by DM patients. The same questionnaire was given to the patient to see whether there is influence questionnaires given to adherence and *self-management*.

Research procedure is used questionnaire BIPQ and EQ-5D. BIPQ contains 9 questions, of which 1 item is a question that is answered by sequencing the main factors causing the disease and 8 other questions representing 8 domains (consequence domain, duration domain of perception, personal control domain, medication control domain, identity domain, domain of concern, domain understanding, and domian emotional response). The quality of life (EQ-5D) questionnaire of the patient is said to be good if the index is close to 1. Visual Analog Scale (VAS) is said to be good if its value is close to 100.

3. Result and discussion

From the research conducted, 80 patients with T2DM with complications meet the inclusion criteria. The results of univariate analysis describe the distribution of respondents including age, sex, social status, education, occupation, income, disease length, length of treatment, and history of disease. Characteristic description of respondents can be seen in Table 1.

IOP Conf. Series: Materials Science and Engineering **259** (2017) 012004 doi:10.1088/1757-899X/259/1/012004

Table 1. Characteristics of patients T2DM Dok II Jayapura hospital

Research Characteristics (n=80)			X±SD
Age (Year)			55.79±10.52
Duration of Treatment			5.25±4.88
Long Disease			6.49±5.46
Research characteristics (n=80)			N(%)
Gender		A. Man	29 (36.3)
		B. Women	51 (63.8)
Education		A. Low education	13 (16.3)
		B. Higher education	67 (83.8)
Work		A. Work	30 (37.5)
		B. Does not work	50 (62.5)
Income		A. Low Income	28 (35.0)
		B. High Income	52 (65.0)
Type of Treatment	Monotherapy:	A. Oral	65 (81.3)
		B. Insulin	
	Combination:	A. Insulin-Insulin	15 (18.8)
		B. Oral-Insulin	
Family history		A. There is	4 (5.0)
		B. There is no	76 (95.0)
Social status		A. Married	77 (96.3)
		B. Single (Not Married, Widowed/Widowed)	3 (3.8)
Comorbidities		A. Microvascular	4 (5.0)
		B. Macrovascular	65 (81.3)
		C. No comorbidities	11 (13.8)

3. 1. Characteristics of respondents Dok II Jayapura

3.1.1. Characteristic description of respondents

In this research, the mean result (mean±SD) of respondent age is 55.79±10.52 year based on the data known to respondent aged 50 years and above more that suffer T2DM. This is in accordance with research [7], where the prevalence of DM will increase with age due to the continued age of insulin expenditure by the pancreas will also be reduced. In another study said that the frequency of diabetics most at the age of 51-60 years [8]. According to previous research, as age increases, the increased incidence of diabetes results from hormonal changes that play a role in metabolism, reproduction and other bodily functions. Female gender (63.8%) was higher than male gender (36.3%) [9]. This is in line with the results of reports of previous studies in which patients with female sex T2DM were more dominant than men [9, 10]. Women are more at risk of diabetes because physically women are more likely to experience increased body mass index. The monthly cycles syndrome (Pre Menstrual Syndrome) and post menopause that make the distribution of body fat become easily accumulated due to hormonal process, so that women are more at risk of suffering T2DM [11]. But according to other studies the data show only the characteristics that do not mean women are more vulnerable to diabetes than men [12], all have the same risk. According to previous research, one of the risk factors of diabetes is heredity, lifestyle (diet, etc.) and age [3].

Based on the last education category it can be seen that majority of respondents have low education level (83.8%) than respondents who have high education level (16.3%). This is in line with studies which suggest that education levels have an effect on the incidence of diabetes mellitus type 2 disease [11, 13]. The higher a person's level of education means there is a better chance of a person's knowledge in preventing the occurrence of illness including T2DM. Characteristics of respondents based on the work found the results of respondents who do not work (62.5%) more dominant than the respondents who work (37.5%). This is because people who do not work plus the elderly and lightweight activities that have a higher risk factor for diabetes mellitus. This is supported by research conducted on 13 European countries it is said that the accumulation of physical activity daily is a major factor determining insulin sensitivity [14]. So the less activity the insulin sensitivity will decrease, the consequence will be a big risk to trigger the occurrence of T2DM. Based on the table, the largest percentage of respondents who suffered from T2DM were respondents with high income (65.0%) compared with respondents with low income (35.0%). The same results were also shown in a study conducted income may be related to the level and lifestyle of a person, so it is feared will affect his health [15]. T2DM therapy in this study was divided into 2 categories: single therapy/monotherapy (therapy of oral and insulin use) and combination (oral-oral, insulin-insulin, and oral-insulin combination). Based on the table obtained the largest (81.3%) of respondents using single therapy (monotherapy) and for the use of therapy in combination obtained the lowest results (18.8%) of respondents using combination therapy. This shows that T2DM patients in Dok II Jayapura Hospital mostly get more monotherapy therapy. The results of the previous study mentioned that patients who suffer T2DM over 5 years will experience various complications of disease that can aggravate the condition of the patient [16].

Based on the results of the study, patients with mean (mean±SD) treatment duration of 5.25±4.88 years. This suggests that T2DM patients in Dok II Jayapura Hospital are mostly those who have long been on diabetes mellitus type 2 treatment. Duration of optimal drug treatment treatment will provide the success of therapy in the treatment of all chronic diseases and improve the quality of life of patients. In diabetes mellitus, adherence to medication affects the success of therapy. Based on the table, which has no family history has the largest percentage (95.0%) and who has a family history has the smallest percentage (5.0%). According to the previous research all have the same risk, one of the risk factors of diabetes is heredity, lifestyle (diet, etc.) [3].

3.1.2. Perception

Assessment of the perception of disease in this study using a perception questionnaire or The Brief-Illness Perception Questionnaire (BIPQ). The highest mean value of perception of disease is found in the Consequence domain (7.63±2.52) and the lowest is in the emotional response domain (3.11±2.19). The average of BIPQ can be seen from Table 2.

Table 2. The average distribution of B-IPQ questionnaires.

Domain	Mean	SD
Consequences	7.63	2.52
Duration Perception	6.50	2.96
Personal Controls	6.20	1.64
Treatment Control	6.73	1.47
Identity	4.80	1.61
Worries	3.34	1.82
Understanding	5.99	1.58
Emotional Response	3.11	2.19

Question 9 is an open question that will illustrate the three main factors causing diabetes mellitus type 2 disease. Based on the results of analysis in this study many factors that menajdi causes diabetes mellitus type 2 disease. After collecting the data and analyzed using descriptive analysis, the most answer to represent 3 main answer cause the first diabetes mellitus type 2 got result equal to 96.3% respondent answer because dietary habit (like sweet food, eat rice, and like oily food and fat) , The second result was 71.3% respondents answered because they like to drink sweet (like sweet drinks like tea, coffee, milk and fast drinks), and the third result is 50.0% of respondents answered because of stress, Thoughts and job matters make a lot of thought and stress. Other descriptive analysis results can be seen in Table 3.

Table 3. Descriptive analysis of the main factors causing T2DM disease

Main factor	Answer 1 (%)	Answer 2 (%)	Answer 3 (%)
Dietary habit	96.3	1.3	-
The drinking pattern	-	71.3	15.0
Stress	-	13.8	50.0

Based on the answers above factors causing diabetes mellitus type 2 disease that has been traced based on the respondents' answers, has been in accordance with the description in [17]. Other factors causing diabetes mellitus type 2 described in [17] are age (with increasing age, the occurrence of glucose intolerance will also increase due to decreased physiological function of the body), history of women who have experienced gestational diabetes mellitus, patients with a history of dyslipidemia, cardiovascular, and patients with *polycystic ovary syndrome* (PCOS).

3.1.3. Quality of life

In this study conducted a univariate analysis which aims to see the picture of the quality of life of T2DM patients using EQ-5D questionnaire. From the Table 4 can be seen that the average quality of life index of patients T2DM 0.920±65.86. The quality of life of the patient is said to be good if the index approaches the value 1. While the mean VAS of patients T2DM is 65.86±9.518. VAS is said to be good if its value is close to 100.

Table 4. The average distribution of the EQ-5D questionnaire.

EQ-5D	Mean	SD
Index	0.920	0.202
VAS	65.86	9.518

3.2. Bivariate analysis

3.2.1. The relationship of respondent characteristics with perception (BIPQ)

In this study bivariate analysis to determine the significance of the relationship characteristics of respondents with perceptions (BIPQ) respondents about the disease. The first step is to test the data normality. In the normality test obtained the results of distribution of data that is not normally distributed. Based on the results of the bivariate analysis in this study using the Mann-Whitney test.
Based on the result of linear regression analysis, there was a significant correlation between age and patient perception about disease seen from p=0.012, p=0.000, p=0.048. The results of this study have been in accordance with previous research, age is one of the characteristics that affect a person's perception of the disease [18]. According to previous research age is an individual characteristic that underlies the occurrence of certain behaviors [19]. The more mature a person will adjust his behavior towards life as well as related to health conditions. Bivariate analysis between relationship of gender characteristic with perception about disease got result that sex genetic characteristic do not have significant relation to entire domain of perception, it seen from p>0.05. Bivariate analysis between the relationship between education characteristic and perception about the disease resulted that the characteristics of education did not have significant relation to the whole domain of perception, it is seen from p> 0.05.

The result of bivariate analysis of work characteristic group with perception about the disease resulted in significant relationship to personal control domain (p=0.003). In the private control domain, the working group of respondents (5.47±1.67) had a better understanding of personal control than the unemployed group (6.64±1.46). Good lifestyle and routine physical activity allow someone to have a better level of health [15]. Perception on the job can affect the perception of the disease, people who do not work tend to have a positive perception because they can perform activities as usual and rarely have complaints related to symptoms of the disease he suffered. The existence of work makes a person busy so they forget about his illness [20, 21]. The result of relation between income with private control domain (p=0.012). In the private control domain, high income groups (6.58±1.47) had a more positive view of the consequences of the disease than the low-income group (5.50±1.73). According to previous research, income will affect the expenditure on health costs. Someone with a low income will be difficult to get treatment because of limited costs, consequently received treatment becomes not optimal [22]. Less optimal treatment can worsen the condition of the patient, because the emergence of various complications that can inhibit daily activities. The existence of financial limitations causes the patient's perception of the disease to be negative. The result of the relationship between treatment type and domain of understanding (p=0.059). In the domain of understanding, the group with the type of monotherapy treatment (5.50±1.73) had a better understanding of the type of monotherapy treatment than the group with the type of combination treatment (6.73±1.43).

The results of this study also states that respondents better understand and more controlled blood sugar using the type of treatment in monotherapy than the type of combination treatment. The statistical results of diabetes in the United States also show similarities, where more patients are taking oral preparations than with insulin. The result of relation between family history with consequence domain (p=0.033), and emotional response (p=0.029). In the group consequences domain that did not have a higher family history (7.75±2.52) had a more positive outlook on the disease than the group with a lower family history (5.25±1.25). In the emotional response domain the group with no family history (3.00±2.17) had a more positive view of the disease than the group with a family history (5.25±1.70). This is in accordance with the study of that individuals with a family history of diabetes consider themselves at greater risk of developing diabetes, compared with those without a family history. The result of a relationship between social status and private control domain (p=0.037). In single person control domains the single group (4.00±1.73) had a more positive view of the disease than the married group (6.29±1.58). This is in accordance with the study of [23], in which respondents with married status are more likely to suffer from diabetes mellitus type 2. A married person has a heavier responsibility in terms of her role as a spouse than an unmarried one. Roles and responsibilities are the weight of this risk factor to the emergence of anxiety effect easy one person experiencing stress.

3.2.2. Relationship characteristics of respondents with quality of life (EQ-5D)

In this study conducted bivariate analyzes to determine the characteristics of the respondents significance relationship with quality of life (EQ-5D) of the respondents about the disease. The first step taken was to test the normality of the data. Normality test results obtained on the distribution of the data were not normally distributed. Based on the results of the bivariate analysis in this study using the test of *Mann-Whitney*.

In this study bivariate analysis to determine the significance of the relationship characteristics of respondents with quality of life (EQ-5D) respondents about the disease. The first step is to test the data normality. In the normality test obtained the results of distribution of data that is not normally distributed. Based on these results bivariate analysis in this study using Mann-Whitney test. Bivariate analysis between age relationship with quality of life found a significant relationship between age with quality of life seen from the value (p=0.007). The results of this study in accordance with previous research, where the prevalence of diabetes mellitus will increase with increasing age then the expenditure of insulin by the pancreas will also be reduced [7]. In another study, T2DM is the most common type of diabetes mellitus, which is about 90-95% of all people with diabetes mellitus and is more commonly experienced by adults over 40 years. This is because insulin retention in diabetes mellitus type 2 incline to increase in the elderly (46-65 years), in addition to the history of obesity and the presence of hereditary factors. Bivariate analysis between the relationship of sex with quality of life found the result that gender does not have significant relationship to quality of life, it is seen from the value p>0.05.

Bivariate analysis between the relationship of education with quality of life found that education has no significant relationship to quality of life, it is seen from the value p>0.05. Bivariate analysis between job relation with quality of life got result that work did not have significant relation to quality of life, it is seen from value $p>0.05$. Bivariate analysis between relation of income with quality of life earn result that earnings have no significant relation to quality of life, it is seen from value $p>0.05$. Bivariate analysis between relation of type of treatment with quality of life got result that type of treatment did not have significant relation to quality of life, it is seen from $p>0.05$. The result of bivariate analysis of group of family history characteristic with quality of life about got result of significant correlation to quality of life. Based on the result (p value=0.025) where (p value <0.05) is in the visual analog scale (VAS) assessment. In the VAS the working group of respondents (75.00±5.77) had a better quality of life against the disease than the unemployed group (65.38±9.45). This is in accordance that individuals with a family history of diabetes consider themselves at greater risk of developing diabetes, compared with those without a family history. The result of bivariate analysis of social status characteristic group found a significant correlation to quality of life. The index (value p=0.000) and VAS (p value=0.001). In the Index Value the married respondents group had a higher quality of life (0.95±0.12) than the single (0.35±0.48) and in the VAS group of married respondents had a higher quality of life (66.96±8.34) than the single group (45.00±5.77). This is in accordance with the study according to[40] suggests that both men and women, individuals with marital status or cohabitation have a higher quality of life. The influence of the marriage status associated with family support is the strongest indicator of having a positive impact on self-care in DM patients [17]. The result of bivariate analysis of the characteristics of comorbidities showed that there was a significant correlation to the quality of life. Based on the result of NI (p value=0.000) and VAS (p value=0.024) where with value $p<0.05$. In the Index Values the respondents of macrovascular concomitant disease had a higher quality of life (0.96±0.11) than microvascular comorbidities (0.69±0.09) and no complications (0.74±0.30) And in the VAS of respondents in macrovascular concomitant disease had a higher quality of life (67.06±8.25) than in the accompanying microvascular disease group (55.00±12.91) and no compilation (62.73±12.72).

This is consistent with studies acute and chronic compilations in DM patients are serious [24], a complication that can occur in patients with type 2 diabetes is ischemic heart disease, stroke and neuropathy. The result of linear regression analysis showed significant correlation between perception of patient to disease with quality of life which can be seen in the perception of medication control value p=0.001, emotional response value p=0.010, understanding p value=0.032, identity value p=0.006, p=0.014 and the duration value p=0.016 . The significance value in the table above showed that the quality of life of T2DM patients will be influenced by perception. This is in accordance with previous study [25] which said that patient perception significantly affects the quality of life and adherence of patients in diabetics.

3.2.3. Perception relationship (BIPQ) with quality of life (EQ-5D)

The result of linear regression analysis showed a significant correlation between patient perception to disease with quality of life which can be seen on live quality domain at index value at treatment control with p=0.000 value and emotional response with p=0.001 and life quality domain VAS value at Understanding with value p=0.032, identity with value p=0.006, at personal control with value p=0.014 and at duration at value p=0.016. So the value of significance shows that the perception of T2DM patients will be influenced by the quality of life.

3.3. Multivariate analysis

3.3.1. Multivariate analysis between the characteristics and the patient's perception (BIPQ)

Based on the Table 5, it can be seen that there are some characteristics that are otherwise significant relationships with all domains on the questionnaire BIPQ. In BIPQ questionnaire reveal any significant differences; perception of the consequences of the characteristics of age, family history, social status and comorbidities; persespsi the duration of the employment, income, family history, social status and duration of disease. Perceptions of personal control of the characteristics of gender, education, occupation, income, duration of treatment, social status, and duration of illness. Perception of control treatment on the characteristics of age, education, family history, social status, and comorbidities. Perceptions of the identity of the characteristics of age and family history. Perception concerns about family history characteristics and comorbidities. Perception understanding of the characteristics of this type of treatment and social status. Perceptions of emotional response to job characteristics, family history, and comorbidities. So that all variables have significant value is selected for inclusion into the multivariate analysis.

Table 5. Spearman correlation test between characteristics and respondents perception (BIPQ).

Variable		Domain of BIPQ							
		C	PD	PC	CT	I	W	U	ER
Age	r	0.298	0.097	-0.035	0.280	-0.197	0.076	0.061	-0.084
	p	0.007 *	0.394	0.758	0.012 *	0.080 *	0.505	0.593	0.460
Type	r	0.115	-0.001	-0.153	0.074	-0.101	-0.098	-0.069	-0.035
Sex	p	0.310	0.996	0.175 *	0.514	0.372	0.388	0.542	0.759
Education	r	0.059	-0.128	0.193	0.166	0.014	-0.068	0.082	-0.074
	p	0.603	0.257	0.086 *	0.142 *	0.900	0.548	0.470	0.515
Job	r	-0.045	-0.172	0.331	0.084	0.098	0.071	0.117	0.187
	p	0.692	0.123 *	0.003 *	0.457	0.387	0.529	0.302	0.097 *
Income	r	-0.107	-0.198	0.282	0.120	0.041	0.112	0.118	0.088
	p	0.345	0.078 *	0.011 *	0.288	0.717	0.323	0.297	0.440
Type	r	-0.055	0.085	0.072	-0.030	0.116	-0.066	0.212	-0.078
Treatment	p	0.627	0.451	0.524	0.794	0.304	0.562	0.059 *	0.490
Long	r	-0.004	-0.087	0.329	0.003	0.056	-0.104	0.066	0.005
Treatment	p	0.974	0.444	0.003 *	0.977	0.622	0.359	0.558	0.967
History	r	0.240	0.178	-0.015	0.215	-0.156	-0.166	-0.134	-0.246
Family	p	0.032 *	0.113 *	0.892	0.055 *	0.167 *	0.141 *	0.238	0.028 *
Status	r	-0.148	0.038	-0.235	-0.207	-0.042	0.084	-0.153	0.120
Social	p	0.191 *	0.737	0.036 *	0.065 *	0.714	0.457	0.175 *	0.288
Comorbidities	r	-0.148	-0.044	0.031	-0.255	0.053	0.164	-0.106	0.163
	p	0.191 *	0.698	0.785	0.023 *	0.639	0.146 *	0.348	0.149 *
Long	r	-0.079	-0.194	0.156	-0.112	0.007	-0.067	-0.039	0.008
Disease	p	0.488	0.085 *	0.167 *	0.322	0.954	0.557	0.732	0.943

Note: C (consequences); PD (Perception duration); PC (personal control); CT (control treatment); I (Identity); W (Worried); U (understanding); ER (emotional response)
P * <0.25

3.3.2. Multivariate analysis on the questionnaire BIPQ

Significant value in the Table 6 shows that the perception of patients with T2DM domains will be influenced by several characteristics. In the perception of the consequences of showing the significant value of 0.012 in age characteristics. In the perception of personal control shows the significant value of 0.002 on job characteristics. In the perception of control treatment showed significant value of 0.000 at age characteristics. In the perception of identity indicates the significant value of 0.048 in life characteristics. In the perception of understanding demonstrated the significant value of 0.043 on the characteristics of the type of treatment. In consequence domain has a significant relationship to the characteristics of age, with p=0.012. This positive correlation of r=0.298 This means the correlation between the consequences with age is directly proportional, meaning that the increasing age of the patient will increase the consequences of the disease in which the increasing age makes one think more of a result that can arise by illness.

IOP Conf. Series: Materials Science and Engineering **259** (2017) 012004 doi:10.1088/1757-899X/259/1/012004

Table 6. Linear regression analysis of the characteristics to perception domain.

BIPQ	Variable	Coefficients		excluded variable		ANOVA		
		Beta	Sign	beta In	Sign	F	Sign	r
Consequence	Age	0.067	0.012	-	-	6.540	0.012	0.298
	Social status	-	-	-0.049	0.669			
Personal Control	Job	1.173	0.002	-	-	10.768	0.002	0.331
	Income	-	-	0.019	0.936			
	Social status	-	-	-0.228	0.032			
	long illness	-	-	0.110	0.304			
Control treatment	Age	0.056	0.000	-	-	14.900	0.000	0.400
	Social status	-	-	-0.124	0.257			
Identity	Age	- 0.034	0.048	-	-	4.035	0.048	- 0.197
Understanding	This type of treatment	0.918	0.043	-	-	4.245	0.043	0.212
	Social status	-	-	-0.187	0.091			

Domain of personal control has a significant relationship to the characteristics of the job, with p=0.002. This positive correlation of r=0.331 means the correlation between personal control at work is directly proportional, meaning that someone who works will improve the patient's personal control of the disease in which a person whose work will be more concerned about themselves against the disease. On the domain control treatment had a significant relationship to the characteristics of age with p=0.000. This positive correlation of r=0.280 means the correlation between the control treatment with age is directly proportional, meaning that increasing the age will improve the treatment of patients towards disease control where increasing age of the patient will make a person more concerned about the treatment of the disease. Domain identity has a significant relationship to the characteristics of age, with p=0.048. The negative correlation seen from the value of this mean r=-0.197 correlation between the control treatment was inversely related to age, which means increasing the age will lower the patient's identity to the increasing age of the disease where it will lower the patient down identify that diabetes mellitus type 2 disease.

In the domain of understanding have a significant relationship to the characteristics of this type of treatment, with p=0.043. This relationship discernible positive correlation of r=0.212 this means understanding the correlation between the type of treatment is directly proportional, meaning that the increased understanding will improve the patient's type of treatment against the disease. According to the research domain can be interpreted as a consequence of an individual's belief about about the severity of the disease and its possible effect on the physical, social and psychological [26]. Good lifestyle and physical activity routine allows one to have a better health level [15].

Domain control describe the beliefs of individual treatment to treatment and expectations of treatment outcome [27]. Identity domain is defined as the idea of the patient's name, their conditions are basically (symptoms related) and the relationship between them [26]. Domain understanding interpreted as associated individuals understanding the disease [28]. Patient understanding of the disease will affect more effective treatment [29].

3.3.3. Multivariate analysis between the characteristics and the patient's quality of life (EQ-5D)

Based on the Table 7 can be seen that there are some characteristics that otherwise there is a significant relationship with the domains on the EQ-5D questionnaire. On EQ-5D questionnaire reveal any significant differences; dimensional index values with the characteristics of age, family history, social status and comorbidities; dimensional visual analog scale with family history and social status. So that all variables have significant value and it selected for inclusion into the multivariate analysis.

Table 7. Spearman correlation test between characteristics of respondents and quality of life.

Variable		Domain EQ-5D	
		Value index	VAS
Age	r	0.181	0.012
	p	0.109 *	0.914
Family history	r	0.142	-0.253
	p	0.208 *	0.024 *
Social status	r	-0.451	-0.326
	p	0.000 *	0.003 *
Comorbidities	r	-0.028	-0.072
	p	0.012 *	0.528

* *Spearman correlate* p <0.25 (Dahlan, 2013)

3.3.4. Multivariate analysis on the EQ-5D questionnaire

Table 8. Linear regression analysis of the characteristics to quality of life domain.

Domain	Characteristics	Coefficients		excluded variable		ANOVA		r
		Beta	Sign	beta In	Sign	F	Sign	
Value index	Age	-	-	0.089	0.275	89.375	0.000	-0.451
	family history	-	-	0.059	0.449			
	Social status	-0.781	0.000	-	-			
	Comorbidities	-	-	-0.179	0.020			
VAS	Family history	-	-	-0.201	0.044	22.132	0.000	-0.326
	Social status	-22.954	0.000	-	-			

Significant value in the table above shows the above table shows that the quality of life of patients with T2DM domains will be affected by a number of characteristics. On the dimension of the index value indicates the significant value of 0.000 on the characteristics of social status, it can be said characteristics variables significant predictors of social status index value. Negative correlation seen from the value r=-0.451, meaning the worse the social status the more it will degrade the quality of life of the index value. Looking at the results of the multivariate analysis can be concluded that the dominant social status can affect the dimensions of the index value in patients with T2DM on quality of life.

The results of the analysis, dimensional visual analog scale shows significant value of 0.000 on the characteristics of social status, it can be said to be a significant characteristic variable of social status become predictors of a visual analog scale. Negative correlation seen from the value r = -0.326, meaning the worse the social status the more it will degrade the quality of life of the visual analog scale. Rate from the multivariate analysis can be concluded that the dominant social status can affect the dimensions of a visual analog scale in patients with T2DM on quality of life. According to research the influence of marital status is related to family support is the most powerful indicator of a positive impact on self-care in patients with diabetes mellitus [30]. These negative attitudes can be overcome with the support of the family. Family support is the most powerful indicator of the positive impact on self-care in patients with diabetes mellitus [30].

3.3.5. Multivariate analysis between perceptions of quality of life

Table 9. Spearman correlation test perceptions of the quality of life domain.

Perception and characteristics		Quality of life value index	VAS
Control treatment	r	0.324	0.073
	p	0.003 *	0.517
Emotional response	r	-0.147	-0.101
	p	0.193 *	0.373
Comprehension	r	0.132	0.361
	p	0.242 *	0.001 *
Identity	r	-0.242	-0.115
	p	0.031 *	0.310
personal control	r	-0.043	0.383
	p	0.703	0.000 *
Duration	r	0.093	-0.349
	p	0.410	0.002 *
Age	r	0.181	0.012
	p	0.109 *	0.914
Job	r	-0.054	0.146
	p	0.632	0.197 *
The type of treatment	r	-0.077	0.120
	p	0.495	0.291

* *Spearman correlate* p <0.25 [31]

Based on the above table it can be seen that there are several domains of perception and characteristics that are otherwise meaningful relationships with quality of life. On the quality of life questionnaire reveal any significant differences; the index value in the control treatment, emotional response, understanding, identity, personal control, duration and age. on a visual analog scale on understanding, personal control, duration and job. So that all variables have significant value and it is selected for inclusion into the multivariate analysis.

3.3.6. Multivariate analysis on the EQ-5D questionnaire

Table 10. Linear regression analysis of perceptions to quality of life domain.

EQ-5D	BIPQ	Coefficients Beta	Coefficients Sign	excluded variable beta In	excluded variable Sign	ANOVA F	ANOVA Sign	r
Value index	Control treatment	0.047	0.001	-	-	9.691	0.000	0.324
	Emotional response	-0.025	0.010	-	-			-0.147
	Comprehension	-	-	0.148	0.166			
	Identity	-	-	0.003	0.979			
	Age	-	-	0.151	0.180			
VAS	Comprehention	1.508	0.038	-	-	11.305	0.000	0.361
	duration	-0.992	0.002	-	-			-0.349
	personal control	1.440	0.039	-	-			0.383
	Job	-	-	-0.077	0.463			

Significant value in the table above shows that the perception of patients with T2DM domains will be influenced by several domains of quality of life. In the domain of life quality index value indicates the significant value of 0.001 in the control domain of treatment. A discernible positive correlation of r=0.324, meaning that the better control of the disease the patient treatment will further improve the quality of life. In the domain of life quality index value indicates the significant value of 0.010 on the domain of emotional response. Negative correlation seen from the value r=-0.147, meaning the patient's negative emotional response to the disease the more it will decrease the quality of life. While the quality of life domain VAS shows the significant value of 0.038 on the domain understanding. A discernible positive correlation of r=0.361, meaning that increasing patient understanding of the disease will improve the quality of life. the quality of life domain VAS shows the significant value of 0.002 on the domain duration. Negative correlation seen from the value r=-0.349, meaning that the longer duration of the disease in patients will increasingly degrade the quality of life of patients. And the quality of life domain VAS shows the significant value of 0.039 on the domain of personal control. A discernible positive correlation of r=0.383, meaning that increasing the personal control of the disease the patient will improve patient quality of life. Looking at the results of the multivariate analysis can be concluded that the control of the dominant treatment and emotional responses can affect dimensional index values and understanding, the duration and the dominant personal control may affect the visual analog scale in patients with T2DM on quality of life.

According to research patient understanding of disease will affect more effective treatment [29]. Perception or understanding of health is influenced by how people believe in its ability to undergo treatment, life, psychosocial, educational owned and family support [32]. Effective treatment will impact the disease uncontrolled type T2DM, so the patient's quality of life can be improved. The existence of a significant relationship to the duration of the domain has been in line with research which mention duration has a significant relationship with quality of life. Patients with chronic diseases who require treatment in the long term will lead to a decrease in adherence when taking the drugs, so that control of the treatment to be bad. Poor control are treatment will reduce the quality of life of patients, due to ineffective treatment can aggravate the patient's disease. In line with previous studies conducted by [33] mentions that there was a significant relationship between perception and quality of life. Patients with chronic diseases are more likely to have unstable emotions (more irritable, anxious and fearful). The response of the negative emotions is what impact the poor quality of life [33].

4. Conclusion

There is a relationship between the perception and the quality of life. Based on the analysis results obtained to improve quality of life of patients with T2DM that is index value in the treatment and control emotional responses as well as on the visual analog scale on understanding, duration and personal control must be changed in order to increase the patient's quality of life. In this case the characteristic factor family history, social status and the type of treatment also affects the quality of life.

References

[1] World Health Organization 2003 The *adherence of long-term therapie Evidence for action* 110 pages [E] ISBN 92 4 154599 2.
[2] RISKESDAS 2013 *Health Research and Development Ministry of Health* Jakarta.
[3] American Diabetes Assocoation (ADA) 2010 The American diabetes association's clinical practice recommendations *Diabetes care* **33** (1) S11-S61.
[4] Mandagi A M 2010 *Factors Relating to the Status of Quality of Life of Patients with Diabetes Mellitus in Puskesmas Pakis Sawahan Surabaya.*
[5] Morisky D E, Ang A, Krousel-Wood M A and Ward H 2008 The predictive validity of medication adherence measure a in an outpatient setting *J. Health-Syst. Pharm.* **10** 348-54.

[6] Thiodora, ynu Gultom 2012 *Level of Knowledge About Diabetes Mellitus Patients with Diabetes Mellitus in the House Management Saki Gatot Subroto Army Central Jakarta* Jakarta Fakultas Ilmu Keperawatan Universitas Indonesia.

[7] Sutikno E 2011 Hubungan antara fungsi keluarga dan kualitas hidup lansia *Medical Journal of Indonesia* **2** 73-9.

[8] Somoza O, Ariana K, Mendoza S and Humberto D 2011 *Diabetes mellitus type 2.*

[9] Adikusumo W 2014 Evaluasi kepatuhan dan kualitas hidup pasien diabetes mellitus tipe 2 yang mendapat monoterapi dan kombinasi terapi antidiabetik oral di rumah sakit umum PKU Muhammadiyah Bantul [*Tesis*] Yogyakarta Universitas Ahmad Dahlan

[10] Stanetic K, Savic S and Racic M 2012 The quality of life of patients with diabetes mellitus type 2 *Onwma meduyuha* **18**(3-4) 7—77.

[11] Irawan D 2010 Prevalensi dan faktor risiko kejadian diabetes mellitus tipe 2 di Daerah Urban Indonesia (analisis data sekunder riskesdas 2007) (Depok: Universitas Press).

[12] Wahyuni Y, Nursiswati and Anna A 2014 Kualitas hidup berdasarkan karakteristik pasien diabetes mellitus tipe 2 Bandung Fakultas Keperawatan Universitas Padjajaran.

[13] Zahatmal C F, Suyatno, and Restuastuti T 2007 The risk factors of diabetes mellitus patients *News Medical Society* **23**(3) 142-147.

[14] Balkau B, Mhamdi L, Oppert J M, Nolan J, Golay A, and Porcellati F 2008 Physical activity and insulin sensitivity *Diabetes* **57** 2613-2618.

[15] Shaikh Z A, Akhund S, Ali M, and Khan M H 2013 Type 2 diabetes Effect of socio-demographic factors among patients *Professional Med. J.* **20** (2) 244-249.

[16] Dinaryanti P, Fudholi A and Andayani T M 2012 Analisis biaya dan efektivitas terapi pasien diabetes mellitus tipe 2 rawat jalan di RSUD Sleman Yogyakarta *Jurnal Manajemen dan Pelayanan Farmasi* **2**(1) 14-19.

[17] PERKENI 2011 *Consensus Management of Type 2 Diabetes Mellitus in Indonesia* Jakarta

[18] Notoatmojdo S 2003 *Education and health behaviors* (Jakarta: PT Rineka Reserved).

[19] Hurlock E B 2004 *Psikologi Perkembangan* (Jakarta: Erlangga).

[20] Picavet H S J and Hoeymans N 2004 Health related quality of life in multiple musculoskeletal diseases : SF-36 and EQ-5D in the DMC3 study *Annals of the Rheumatic Diseases* **63** 723-729.

[21] Broadbent E, Petrie K J, Main J, and Weinman J 2006 The brief illness perception questionnaire (BIPQ) *J. Psychosom. Res.***60** 631-637.

[22] Notoatmojdo S 2010 *Health promotion, theory and applications* (Jakarta: Rineka Cipta).

[23] Al-Maskari M Y, Al-Shooki A O, Al-Adawi S H and Lin K G 2011 Assesment of quality of life in patients with type 2 diabetes mellitus in oman *Saudi Med. J.* **23**(12) 1285-1290.

[24] Solli O, Stavem K, and Kristiansen I S 2010 Health related quality of life in diabetes: The associations of comlications with EQ-5D score *Health And Quality of Life Outcomes* **8**(18) 1-8.

[25] van Erp S J H, Brakenhoff L K P M, Vollmann M, van der Heijde D M, Veenendaal R A, Fidder H H, Hommes D W, Kaptein A A, van der Meulen-de Jong A E and Scharloo M 2017 Illness perception and outcomes in patients with inflammatory bowel disease: coping is a mediator? *Int. J. Behav. Med.* **24**(2) 205 – 2014.

[26] Weinman J 1996 *The ilness perception questionnaire: a new method for assesing the cognitive representation of illness* (Netherland: Harwood Academic Publisher).

[27] Moss-Morris R, Weinman J, Petrie K J, Horne R, Cameron l D, and Buick D 2002 The revised illness perception questionnaire (IPQ-R) *Psychology and Health* **17** 1-16 .

[28] Taylor S E 2009 Health Psychology Seven Edition (USA: McGraw-Hill).

[29] Petrie K J, Jago L A, Devcich D A 2007 The role of illness perception inpatients with medical conditions *Current Opinion in Psychiatry* **20** 163-167.

[30] Hensarling J 2009 *Development and testing of Hensarling diabetes psichometric family support scale* Dissertation Degree of Doctor of Philosophy in the Graduate school Woman's University Texas.

[31] Dahlan S M 2013 *Besar Sampel dan Cara Pengambilan Sampel* (Jakarta: Salemba Medika).
[32] Pricahyo H D 2012 Perception differences male patient compliance on diet and diabetes mellitus in Space Rose B dr. Harjono Ponorogo [*Essay*] Prodi DII Nursing Faculty of Health Sciences University of Muhammadiyah Ponorogo.
[33] Wahl A K, Robinson H S, Langeland E, Larsen M H, Krogstad A L, and Moum T 2013 Clinical characteristics associated with illness perception in Psoriasis *Acta Dermato-Venereologica* **93**.

Physical characteristic and irritation index of *Syzigium aromaticum* essential oil in O/W and W/O creams

R Safriani, N Sugihartini, S Yuliani

Faculty of Pharmacy, Universitas Ahmad Dahlan, Yogyakarta, Indonesia

E-mail: nining.sugihartini@pharm.uad.ac.id

Abstract. Essential oil of *Syzigium aromaticum* has been formulated in O/W and W/O creams as anti-inflammatory dosage form. The purpose of this study was to evaluate the physical characteristic and irritation index of *S. aromaticum* essential oil in O/W and W/O creams. The creams were made by fusion method. The creams then were evaluated the physical characteristic including pH, viscosity, spreadability and adhesivity. The irritation index was obtained by irritation skin test in male rabbit. The results showed that the W/O and O/W creams have the value of pH: 6.3 and 6.27; spreadability: 3,18 and 4.17 cm^2; adhesivity: 5.59 and 0.07 minutes; viscosity: 4.43 and 2.88 Pa.S, respectively. The irritation test showed that the control enhancer caused mild irritation in both of W/O and O/W creams. These findings indicated that type of cream might influence the physical characteristic and irritation index of *S. aromaticum* essential oil cream.

Keywords: *S. aromaticum*, essential oil, physical characteristic, irritation index, cream

1. Introduction

The essential oil of *S. aromaticum* has efficacy as anti-inflammatory. The previous study have formulated the essential oils in various types of ointment base. The results of this study indicated that the difference in the composition of the formula affected the physical properties of the dosage forms. Data of adhesivity indicated that the increase of essential oil concentration leaded to an increase in the adhesivity of the oinment with base of W/O cream, lotion, hydrocarbons and the decrease of the adhesivity of the oinment with absorption base. Data of spreadability showed that the increase of the essential oils concentration caused an increase the spreadability of oinment with absorption W/O and O/W creams, and hydrocarbons bases, while on the lotion base there was a decrease of the spreadability. The spreadability of emulgel base was not affected by the increase of concentration of essential oil of *S. aromaticum*. Data of pH showed that the increase of essential oil concentration of *S. aromaticum* did not affect the pH in absorption base, W/O cream, lotion, and hydrocarbon. An increase in pH occur on the O/W cream with the an increase of the concentration of essential oils [1-7]. Other studies have also shown similar results. The formulation of ethanol extract of mangosteen peel in gel showed that variation of gelling agent would affect the adhesivity and spreadability of gel. The results of study showed that carbopol was the most optimal gelling agent in the formulation. The increase of concentration of ethanol extract in gel raised the adhesivity of gel and capability of gel as wound healing. The other study showed that composition variation of oleic acid and propylene glycol as enhancer affected the activity of the essential oil of *S. aromaticum* as anti-inflammatory. The increase of concentration of propylene glycol caused the decrease of the number of cells with COX-2 expression, the number of inflammatory cells and the epidermal thickness of skin [8-10].

In this study the essential oil of *S. aromaticum* was formulated in W/O and O/W creams with the addition of enhancer. The enhancer composition of O/W cream was 50% oleic acid and 50% propylene glycol. While the enhancer composition on the W/O cream was 30% oleic acid and 70% propylene glycol. The results of previous study showed that the difference of composition in the formula woud affect the physical properties of the preparation. Thus, the determination of the physical properties and the irritation index of essential oil of *S. aromaticum* formulated in W/O and O/W creams were needed.

2. Materials and Methods

2.1. Materials

This study used essential oil of *S. aromaticum* from the Center for Essential Oils Studies Islamic University of Indonesia (CEOS-UII), Yogyakarta, Indonesia. The ingredients for the O/W and W/O creams were pharmaceutical grade i.e white vaseline, methyl paraben, prophyl paraben, propylene glycol, stearic alcohol, natrium lauryl sulphate, aquadest, oleic acid, cetaceum, cera alba, parafin liquidum and natrium tetra borate. Rabbit was used as an animals in irritation skin test. The whole procedure in this study was approved by the Research Ethics Committee of Universitas Ahmad Dahlan with approval number No. 011504040.

2.2. Methods

2.2.1. Formulation of O/W and W/O creams

The O/W and W/O type creams were made using fusion method. Each of the oil phase and the water phase was heated at 70^0C. The ingredients of oil phase of O/W cream were white vaseline, prophyl paraben, stearic alcohol, natrium lauryl sulphate, oleic acid, whereas in cream type W/O were cetaceum, cera alba, parfin liquidum, oleic acid. The ingredients of water phase of O/W cream were propylene glycol, methyl paraben, aquadest whereas w/o type cream were natrium tetra borate, propylene glycol, and aquadest. After that, the oil and the water phases of each cream were mixed until homogeneous. The essential oil of *S. aromaticum* was added after the base of cream getting cold [11, 12].

2.2.2. Evaluation of physical characteristic

2.2.2.1. Adhesivity test

Briefly, 0.25 g of cream was weighed and put between two glass objects. Then, 1 kg of load has placed upper the glass objects to give a pressure for 5 minutes. After that, the glass objects were placed on the tool that was given load 80 g. The time needed to separate the two glass objects after the load was released was recorded [13].

2.2.2.2. Spreadability test

Briefly, 0.5 g of the cream was weighed and put on the midle of circular glass. The other glass was put on the upper of it for 1 minute. The diameter of cream was measured. The 100 g of load was put on the glass for 1 minute and then the diameter of cream was measured until getting the constant one[14].

2.2.2.3. pH test

Five hundred mg of cream was diluted with 5 ml of distilled water. The pH of cream was then measured using pH meter [15].

2.2.2.4. Viscosity test

The viscosity of cream was measured by using viscosimeter Rheosys Merlin VR

2.2.3. Irritation skin test

This test refered to the BPOM guidelines for non-clinical toxicity test (*in vivo*). The test used six male rabbits 3 months old. Firstly, the back hair of rabbits were shaved carefully, then the removal cream of hair was applied. After 48 hours, the backs of rabbits were divided into 4 square. It was used to apply

IOP Conf. Series: Materials Science and Engineering **259** (2017) 012005 doi:10.1088/1757-899X/259/1/012005

the O/W type cream, W/O type cream and crotton oil 4% (as negatif control). One square was used as as normal control. The observations were conducted for 24, 48 and 72 hours. To study the reversibility, the observation was continued after 7 and 14 days. The responses of the cream application were assessed according to the guidelines [16].

3. Results and Discussion

3.1. The physical characteristic of cream

The physical characteristic of O/W and W/O type creams were presented in Table 1.

Table 1. Physical characterististic of O/W and W/O creams

Parameter	O/W type cream x±sd (n=3)	W/O type cream x±sd (n=3)
Viscosity (Pa.s)	4.43±0.15	2.88±0.66
Spreadability (cm^2)	3.18±0.10	4.17±0.12
Adhesivity (minutes)	5.59±0.35	0.07±0.009
pH	6.63±0.08	6.27±0.08

The viscosity of O/W cream was greater than W/O cream. This was due to differences in the composition of the ingredients in each cream. The one of component of the O/W type cream was stearyl alcohol which can provide good viscosity and adhesion [17]. The study on the formulation of several active ingredients in O/W, W/O and amphifilic type creams also showed that O/W cream had the highest viscosity compared to other types of cream [18].

In addition, the presence of liquid paraffin in W/O type cream also caused the consistency of cream to be softer causing the viscosity becomes smaller [15]. The differences in viscosity may due to differences in composition of enhancer. The enhancer compositions were 50% oleic acid: 50% propylene glycol in O/W type cream and 30% oleic acid:70% propylene glycol in W/O cream. The amount of propylene glycol in W/O type cream was greater than in O/W type cream. The function of propylene glycol was as humectan which caused the w/o type cream had lower consistency.

The viscosity of the cream will affect the spreadability and adhesivity of the cream. The higher viscosity caused the lower of consistency and resulted in the decrease of spreadability. On the other hand, it caused increase of adhesivity. This was also in accordance with the results of previous studies on essential oil of *S. aromaticum* formulations in the base of absorption and base of lotion [1,6]. Based on the previous study, the viscosity of O/W type cream higher than other type cream was caused by the stearyl alcohol in the formulation [17]. The spreadability of cream described the ease of cream in covering the wound. Therefore cream preparations are expected to have a wide spread. Good spreading cream was about 5-7 cm^2 [19]. The adhesivity described the ability of the cream to adhere to the wounded skin. Therefore, cream was expected to have long adhesive power. The cream should have at least 4 seconds of adhesion [20].

Testing of pH was intended to know the level of acidity of cream which not cause skin irritation [21]. The result of pH test showed that the pH of the O/W and W/O type creams were in the range of skin pH (4.5 to 6.5). Topical preparations were expected to have a pH that was in the range of normal skin. The pH of cream was upper the pH of normal skin will trigger scaly skin, and the pH of cream was under the pH of normal skin will trigger skin irritation [22]. The pH of O/W cream was greater than that of W/O type cream. The amount of oleic acid as enhancer in O/W cream was greater than in W/O cream. The composition of enhancers were 50% oleic acid:50% propylene glycol in O/W type cream and 30% oleic acid:70% propylene glycol in w/o type cream.

3.2. The irritation test

The result of irritation test was presented in Table 2.

Table 2. The index irritation of creams

Group of test	W/O cream	O/W cream
Normal control	0	0
Base of cream	0	0
Base of cream + Enhancer	0	0
Base of cream + Enhancer + Essential oil of *S. aromaticum*	0	0.03
Base of cream + Essential oil of *S. aromaticum*	0	0
Enhancer	0.13	0.13

Data in Table 2 showed that enhancer caused mild irritation both in W/O and O/W creams. Combination of oleic acid and propylene glycol was used as enhancer in this study. Enhancer have mechanismed by damaging the lipid tissue so it can cause a certain side effects on the skin [23]. The formulation contained each of enhancer and essential oil of *S. aromaticum* did not cause irritation both in W/O and O/W type creams. It mean the formulation could reduce side effect of enhancer. On the other hand, the formulation contained mixture of enhancer and essential oil of *S. aromaticum* in o/w cream caused mild irritation. It may due the amount of enhancer and eugenol in essential oil of *S. aromaticum* was released higher than in w/o cream. The composition of water phase in O/W type cream was higher than in w/o type cream. In this condition, the non polar component such as oleic acid and essential oil of *S. aromaticum* were easier to release from O/W cream than from type w/o cream. There are two components that can cause mild irritation i.e enhancer and eugenol in essential oil of *S. aromaticum*. Based on the literature, essential oil might cause mild irritation[24]. Eugenol is an active compound in *S. aromaticum* essential oil which have enhancer activity [25]. Essential oil could be used as enhancer because it contains of terpenes that can be used as enhancers [26-28]. Terpen has high ability to improve the penetration and low systemic toxicity at low concentrations (1-5%). It can increase the polar drug's diffusion coefficient in the membrane by altering the fat structure of the stratum corneum [29].

4. Conclusions

Type of cream may influence the physical characteristic and irritation index of *S. aromaticum* essential oil cream.

Acknowledgement

This study was supported by grant Hibah Tim Pascasarjana from DIKTI in 2017 No. 118/SP2H/LT/DRPM/IV/2017

References

[1] Vicky A K, Sugihartini N and Yuwono T 2016 Physical properties and irritation index essential oil of clove (*Syzygium aromaticum*) in absorption base ointment with variation concentration *Proc Int Conf of CONFAST 2016 Conference series: International Conference on Industrial Biology* (Yogyakarta) **1746**(1) (AIP Conference) 10.1063.

[2] Haque A F, Sugihartini N, Yuwono T 2015 Evaluasi uji iritasi dan uji sifat fisik pada sediaan krim m/a minyak atsiri bunga cengkeh (*Syzigium aromaticum*) dengan berbagai variasi konsentrasi *Pharmacy* **12**(2) 131.

[3] Sari D K, Sugihartini N and Yuwono T 2015 Evaluasi uji iritasi dan uji sifat fisik sediaan emulgel minyak atsiri bunga cengkeh *(Syziqium aromaticum) Pharmaciana* **5**(2) 115.

[4] Pratimasari D, Sugihartini N and Yuwono T 2015 Evaluasi sifat fisik dan uji iritasi sediaan salep minyak atsiri bunga cengkeh *(Syziqium aromaticum)* dalam basis larut air *JIF* **11**(1) 9.

[5] Pranawati E, Sugihartini N and Yuwono T 2016 Sifat fisik dan daya iritasi krim tipe a/m minyak atsiri bunga cengkeh *(Syziqium aromaticum)* dengan berbagai variasi konsentrasi *JIF* **12**(1) 1.
[6] Latifah F, Sugihartini N and Yuwono T 2016 Evaluation of physical properties and irritation index of lotion containing *Syzygium aromaticum* clove essential oil at variation concentration *Trad. Med. J.* **21**(1) 1.
[7] Mukhlishah N R I, Sugihartini N and Yuwono T 2016 Daya iritasi dan sifat fisik sediaan salep minyak atsiri bunga cengkeh *(Syziqium aromaticum)* pada basis hidrokarbon *Majalah Farmaseutik* **12**(1) 372.
[8] Fujiastuti T and Sugihartini N 2015 Sifat fisik dan daya iritasi gel ekstrak etanol herba pegagan (*Centella asiatica*) dengan variasi jenis gelling agent *Pharmacy* **12**(1) 11.
[9] Sugihartini N and Wiradhika R Y 2017 Gel formulation of ethanol extract of mangosteen peel (*Garcinia mangostana*, L) as a medication for burns in wistar rats *JKKI* **8** (2) 110.
[10] Iriani FA, Sugihartini N and Yuwono T 2017 The profile of anti-inflammatory activity of *Syziqium aromaticum* volatile oil in lotion with variation composition of oleic acid and propylene glycol as enhancer *Trad. Med. J.* **22**(2) 111.
[11] Yuliastuti D 2016 optimasi komposisi *enhancer* asam oleat dan propilen glikol krim minyak atsiri bunga cengkeh tipe M/A dengan metode simplex lattice design *Thesis* Magister Program of Pharmacy Universitas Ahmad Dahlan.
[12] Tuldjanah M, 2016 Optimasi komposisi *enhancer* asam oleat dan propilen glikol pada sediaan krim tipe A/M minyak atsiri bunga cengkeh sebagai antiinflamasi dengan metode *simplex lattice design Thesis* Magister Program of Pharmacy Universitas Ahmad Dahlan.
[13] Rahmawati D, Sukmawati A and Indrayudha P 2010 Formulasi krim minyak atsiri rimpang temu giring (*Curcuma heyneana* Val & Zijp): Uji sifat fisik dan daya antijamur terhadap *Candida albicans* secara in vitro *Trad. Med. J.* **11**(2) 137.
[14] Astuti I Y, Hartanti D and Aminiati A 2010 Peningkatan aktivitas antijamur *Candida albicans* salep minyak atsiri daun sirih (*Piper bettle* LINN.) melalui pembentukan kompleks inklusi dengan β-siklodekstrin *Trad. Med. J.* **15** 94.
[15] Naibaho O H, Paulina V Y, Yam Lean and Wiyono W 2013 Pengaruh basis salep terhadap formulasi sediaan salep ekstrak daun kemangi (*Ocinum sanchum* L.) pada kulit punggung kelinci yang dibuat infeksi *Staphylococcus aureus Pharmacon Jurnal Ilmiah Farmasi-UNSRAT* **2**(2) 27.
[16] Anonim 2014 *Pedoman Uji Toksisitas Non-Klinik Secara In-Vivo* Badan Pengawas Obat dan Makanan Republik Indonesia, Jakarta) 1.
[17] Kulawik-Pioro A and Potykanowicz A 2016 Determining the quality of hydrophobic barrier creams by rheological measurements sensory analysis, ph determination and permeation time measurements *Chemometrics and Intelligent Laboratory Systems* **156** 14.
[18] Nagelreiter C, Kratochvilova E and Valenta C 2015 Dilution of semi-solid creams: influenced of various production parameters on rheological properties and skin penetration *Int. J. Pharm.* **478** 429.
[19] Garg A, Aggarwal I D, Garg S and Sigla A K 2002 Spreading of semisolid formulation *An Update, Pharmaceutical Technology* 84.
[20] Ulaen P J, Selfie, Banne, Suatan Y and Ririn A 2012 Pembuatan salep anti jerawat dari ekstrak rimpang temulawak *(Curcuma xanthorrhiza Roxb.) JIF* **3**(2)45.
[21] Mappa T, Edi J H and Kojong M 2013 Formulasi gel ekstrak daun sasaladahan (*Pperomia pellucida* L.) dan uji efektivitasnya terhadap luka bakar pada kelinci *JIF* **2**(20) 49.
[22] Swastika A, Mufrod and Purwanto 2013 Aktivitas antioksidan krim ekstrak sari tomat (*Solanum lycopersicum L.) Trad Med Journal* **18** (3)132.
[23] Fox L, Gerber M, Plessis J, and Hamman J 2011 Transdermal drug delivery enhancement by compound of natural origin *Molecules* **16** 10507.
[24] Vanish I, Ahad A, Aqil M and Agarwal SP 2014 Investigating the potential of essential oils as

penetration enhancer for transdermal losartan delivery: effectiveness and mechanism of action *AJPS* **9** 260.

[25] Sheth N S and Mistry R B 2011 Formulation and evaluation of transdermal patches and to study permeation enhancement effect of eugenol *JAPS* **01**(03) 96.

[26] Wang J, Lan Y, Li H, Zhang Y, Zhang Q, Cao Y and Wu Q 2014 Percutaneous penetration enhancement effect of essential oil of mint (*Mentha haplocalyx* Briq) on Chinese herbal components with different lipophilicity *J. of Trad. Chinese Med. Sci.* **1** 109.

[27] Tak J H and Isman M B 2017 Enhanced cuticular penetration as the mechanism of synergy for the major constituents of thyme essential oil in the cabbage looper, Trichoplusia ni *Ind. Crops. Prod.* **101** 29.

[28] Barradas T N, Senna J P, Cardoso S A, Nicoli S, Padula C and Santi P 2017 Hydrogel-thickened nanoemulsions based on essential oils for topical delivery of on essential oils for topical deliveryof psoralen: permeation and stability studies *Eur J. Pharm. Biopharm.* **116** 38.

[29] Kanikkanan N, Kandimalla K, Lamba S and Singh M 1999 Structure-activity relationship of chemical penetration enhancers in transdermal drug delivery *Curr. Med. Chem.* **6**(7) 593.

The cytotoxic effect of *Elephantopus scaber* Linn extract against breast cancer (T47D) cells

N Sulistyani, Nurkhasanah

Faculty of Pharmacy, Universitas Ahmad Dahlan, Yogyakarta, Indonesia

E-mail: nurkhas@gmail.com

Abstract. Breast cancer is one of the main cause of death. *Elephantopus scaber* Linn (ES) which has been used as a traditional medicine contains an antitumor compounds. This study aimed to explore the active fraction from ethanolic extract of ES as anticancer and to determine its inhibition effect on the cell proliferation cycle of breast cancer (T47D) cells. The ES leaf was macerated with ethanol and then evaporated to get the concentrated extract. The extract was fractionated using petroleum ether, chloroform, and methanol respectively. The cytotoxic activity of each fraction was carried out with MTT method, and the inhibition of cell cycle test were observed by flowcytometry method. The result showed that ES and the fractions have cytotoxic activity against T47D cell lines with IC_{50} values of extract, petroleum ether, chloroform, and methanol fractions were 58.36±2.38, 132.17±9.69, 7.08±2.11, and 572.89±69.23 μg/mL. The inhibition effect of ethanol extract on the lifecycle of cells was occured in sub G1 phase. There was no prolonging of G1, S, G2/M and polyploidy phase of T47D cell lines. The chloroform fraction of ES is the most cytotoxic fraction against T47D cells without prolonging the cell lifecycle.

Keywords: *Elephantopus scaber*, cytotoxic, T47D, IC_{50}, cell cycle

1. Introduction

Breast cancer is the most common cancer among women and often causes death. The incidence of breast cancer is still high and difficult to be cured although the therapy has developed and progressed rapidly. The common treatment of breast cancer is consist of surgery, radiotherapy, chemotherapy and hormone therapy [1, 2].

Chemotherapy is a choise to stop the growth of cancer cell, but it is very toxic and has many side effects. The failure of chemotherapy can be associated with the failure of anticancer agents to induce programmed cell death (apoptosis). There are many reports of cancer cells resistance to chemotherapy. The resistance can be caused by overexpression of PGP in cells that lead to the presence of drug efflux out of the cell. Therefore, development of new cytotoxic agents for cancer therapy is urgently needed [3].

Many cytotoxic agents were provided in medicinal plants. *Elephantopus scaber* is a plant which has been reported as cytotoxic agent. The extract and fractions of *E. scaber* was proven to induce apoptosis against cervical cancer [4, 5]. *E. scaber* has been used traditionally to treat various deseases [6].

The anticancer potency of some chemotherapy was associated with the capability to inhibit the cells growth at certain phase of cell lifecycle. Based on differences in DNA content, cells can be distributed in phases of the cell cycle i.e. sub G1, G1, S, G2 and M as well as the polyploidy cells [7, 8]. The objective of this research was to explore the potency of ES extract as cytotoxic agent against breast cancer cells and screening the active fraction from the extract. The active fraction was then assayed the potency for inhibiting cell cycle.

2. Materials and Methods

2.1. Plant material collection

The plant material was collected from Merapi Farma, Yogyakarta, Indonesia. It was dried and blended.

2.2. Extraction and fractionation

The 200 g of ES powder was macerated using 900 mL of ethanol. The extract was collected and evaporated to get the concentrated extract. The concentrated extract was subsequently dissolved in petroleum ether and shaked for 6 hours and allowed to equilibrium for 24 hours. The soluble fraction was separated as petroleum ether fraction and the non soluble fraction was then fractionated using chloroform, ethyl acetate and methanol respectively. All fractions were evaporated and the dried fractions were collected.

2.3. Cytotoxicity test

Cytotoxicity assay was carried out using MTT method. The 100 μL suspension of T47D cells at a density of $1x10^4$ in 96 wells micro-plate were incubated for 24 h at 37°C (5% CO_2) with 100 μL RPMI medium and ES fraction with concentration series of test 2000; 1500; 1000; 800; 400; 200; 100; 50; 25; 12.5; 6:25; and 3.125 μg/mL.

After that, media was discarded by inverting the plate slowly ad followed by adding 100 μL of MTT, then incubated for 4 hours at 37°C, 5% CO_2. After incubation, the mixture were added with 100 mL of SDS solution in 0.01 N HCl and incubated for over night at room temperature. The microplate were read by ELISA reader at 550 nm wavelength. The IC_{50} values were then calculated from percentage of living cells [9].

2.4. Flowcytometric analysis

The T47D cells was treated using ES extract with concentration equal to $½IC_{50}$ and IC_{50} and incubated for 24 hours. The cells was then washed using PBS and subsequently was added with 100 μL of Annexin V-PI and 350 μL of PBS. Furthermore, the cells was given DNase free RNase (20 μg/mL) and incubated at 37°C for 10 m. The cell was analysed using FACS Calibour Flowcytometry.

3. Results and Discussion

Medicinal plants produce a lot of bioactive compounds and can be used in treatment of many deseases. The exploration of therapeutic potential of the plants has been carried out since several decades ago. One of them is addressed to find the potential of anticancer. Cancer is one of the major health problems and lead to a high of death numbers. Breast cancer is the most common cancer among women. The great potential of plant-based compounds for the treatment and prevention of cancer is attributed to their safety, low cost, and oral bioavailability.

E. scaber have been used traditionally to treat various diseases as antiinflammation, diarrhea, hepatitis, arthritis. With the diverse traditional applications of *E. scaber*, United Nations Development Program has recommended *E. scaber* as a potential natural herb which should be further studied [6].

3.1. Cytotoxicity test

Fractionation process was conducted using petroleum ether, chloroform and methanol to separate the compounds with different polarity. It would be useful to get the fraction with the highest cytotoxicity. In this study, T47D cell lines were used as targetted cell in the cytotoxicity test of ES fractions. The T47D cell lines was derivated from breast cancer cells, so it can become the tested cells to find the potential activity of ES fraction against breast cancer cells. The citotoxic activity was evaluated by MTT method [10]. It was based on the activity of tetrazolium succinate reductase enzyme in viable cells reducing tetrazolium salt *(3-(4,5-dimetiltiazol-2-il)-2,5-difeniltetrazolium bromide)* to produce formazan crystal [9]. The concentration of dissolved formazan was measured by spectrophotometric method.

Figure 1 showed the percentage of T47D cell viability after ES fraction treatment. The viability of T47D cells were shown in dose dependent manner. The higher concentration resulted the more viability of the cells. The IC_{50} value of ES extract and each fraction were performed in Table 1.

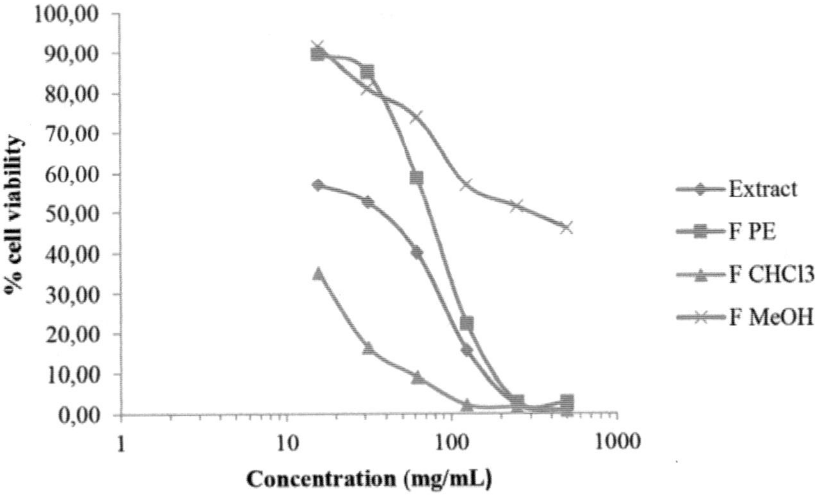

Figure 1. The cytotoxicity of ES extract and fractions against T47D cells.

Table 1. The cytotoxicity of ES extract and fraction against T47D cells presented as IC_{50}.

Fraction	IC_{50} (μg/mL)
Extract	58.36 ± 2.38
F PE	132.17 ± 9.69
F $CHCl_3$	7.08 ±2.11
F MeOH	572.89 ± 69.23

Based on the Table 1, it's revealed that chloroform fraction (F $CHCl_3$) treatment resulted the lowest IC_{50} indicated the high cytotoxicity against T47D cells. Some chemical constituent has been reported and identified as antitumor were deoxyelephantopin [11], scabertopinol, trans-caffeic acid, methyl 3,4-dicaffeoylquinate, luteolin-41-O-β-D-glucoside, trans-p-coumaric acid, indole-3-carbaldehyde, methyl trans-caffeate, luteolin-7-O- glucuronide 6"-methyl ester, and luteolin [12].

Figure 2 showed the cytotoxicity of the chloroform fraction. The morphology of cells in the control group (without ES fraction treatment) is very different from the treatment (using 1.5625 μg/mL of chloroform fraction) group. In the control group, the cells shape is tapering with nucleus looked clearly. On the other hand, the cells in the teatment seemed more little and circle with lower density than control.

IOP Conf. Series: Materials Science and Engineering **259** (2017) 012006 doi:10.1088/1757-899X/259/1/012006

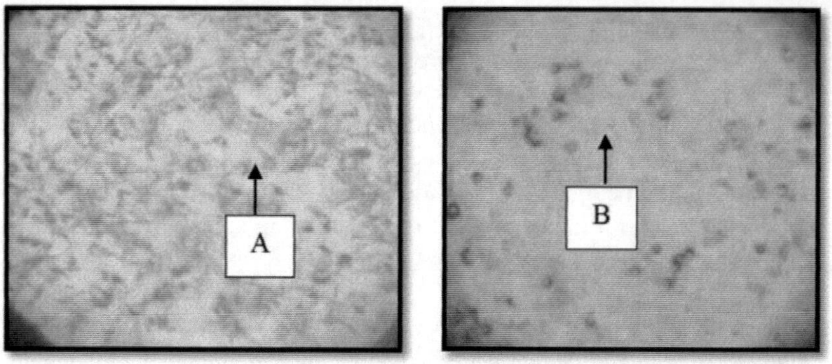

Figure 2. Microscopic performance of T47-D cells of control group (A) and group of treatment with 1.5625 μg/mL of chloroform fraction (B)

3.2. Flowcytometric analysis

The flowcytometric analyses was subject to study the effect of ES extract to the cell cycle. The flowcytometry can inform the cells distribution in phases of sub G1, G1, S, G2/M, and the polyploidy cells based on the amount of chromosome set. Effect of the extract can be known by comparing the effect between control and treatment. The concentrations of ES extract used in this study were equal to ½ IC_{50} and IC_{50}.

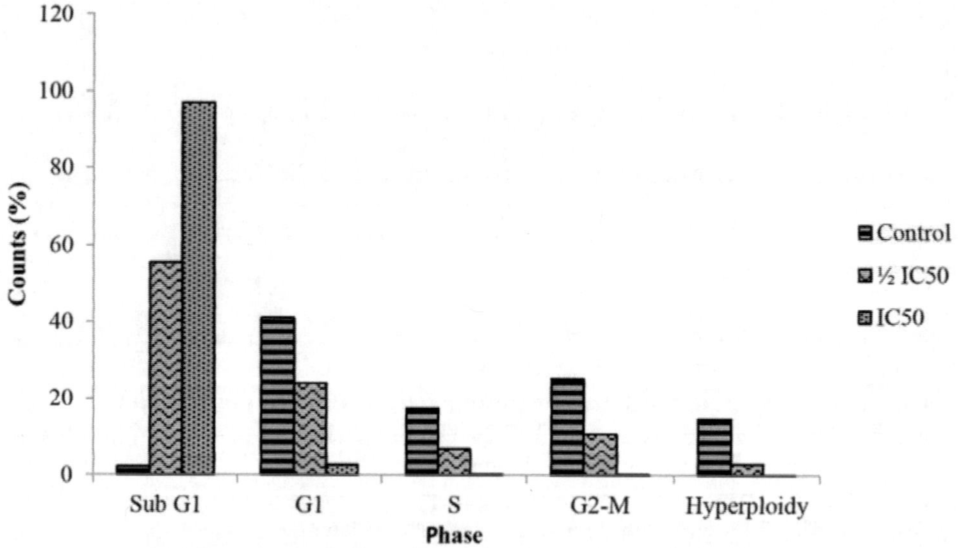

Figure 3. Comparation of T47D cell lines distribution in 5 cell phases among the control and the treatment with ½ IC_{50} as well as IC_{50} concentration of ES extract.

Figure 3 showed that in the control, the cells were distributed in all phases, mainly in G1 (41.07%), S (17.34%), and G2-M (25.34%) phases. The sub G1 phase only contained 2,27% of cells total. In the treatment of ½ IC_{50} of ES extract, the cells were most accumulated in Sub-G1 phase (55.34% of cells total). There were decending of cell concentration in phase of G1, S, G2-M and the polyploidy cells compared to the control. It means that this treatment resulted cell growth inhibition since the sub G1 phase or the cells were dead before growing to the next stage. Nevertheles, the other cells still develop in the next steps. The same result occured when the cells were treated by ES extract with the concentration equal to IC_{50}. Almost all of cells (96.95% of cells total) were detected in sub G1 phase. This ES extract concentration arrested all cells in sub G1. Accumulation of cells in sub G1 phase may consist of apoptotic and/or necrotic cells [13–15]. Many factors may contribute the apoptosis as well as necrosis process. Hence, this study can be use to observe what mechanism inducing cells death in the next study.

4. Conclusions

E. scaber extract exerted anti-cancer activity on breast cancer (T47D) by cell cycle arrest in sub G1 phase. The chloroform fraction of *E. scaber* ethanolic extract was the highest active fraction.

Acknowledgement

The authors thank to Ministry of Research, Technology and Higher Education for funding this research. We also thank to Bariadi Nugroho, Panji Ratih Suci, Emi Syafaati, Kuntum Zahro Wardati, Irma Sari Tilawah and Muhammad Lukman for assisting during research.

References

[1] Sharma G N, Dave R, Sanadya J, Sharma P and Sharma K K 2010 Various types and management of breast cancer: An overview *J. Adv. Pharm. Technol. Res.* **1**109–26

[2] Siegel R, DeSantis C, Virgo K, Stein K, Mariotto A and Smith T 2012 Cancer treatment and survivorship statistics 2012 CA Cancer *J. Clin.* **62** 220–41.

[3] Reddy L, Odhav B and Bhoola K D 2003 Natural products for cancer prevention: a global perspective *Pharmacol. Ther.* **99** 1–13.

[4] Beeran A A, Maliyakkal N, Rao C M and Udupa N 2015 The enriched fraction of Elephantopus scaber Triggers apoptosis and inhibits multi-drug resistance transporters in human epithelial cancer cells *Pharmacogn. Mag.* **11** 257–68.

[5] Nurkhasanah, Trisnamurti K C, Gunaryanti R D, Widyastuti T and Listyowati Y 2015 The screening of cytotoxic fraction from *Elephantopus scaber* Linn against human cervical cancer (Hela) Cells *Int. J. Pharma. Sci. Res.* **6**(6) 1011-14.

[6] Ho W Y, Ky H, Yeap S K, Rahim R A, Omar A R and Ho C L 2009 Traditional practice , bioactivities and commercialization potential of *Elephantopus scaber* Linn *J. Med. Plant. Res.* **3** 1212–21.

[7] Darzynkiewicz Z, Halicka H D and Zhao H 2010 Analysis of cellular DNA content by flow and laser scanning cytometry *Adv. Exp. Med. Biol.* **676** 137–47.

[8] Roh M, van der Meer R and Abdulkadir S A 2012 Tumorigenic polyploid cells contain elevated ROS and ARE selectively targeted by antioxidant treatment. *J. Cell. Physiol.* **227** 801–12.

[9] Doyle A, Griffiths S J B 2007 *Cell And Tissue Culture For Medical Researc*h 1 edition (New York: John Wiley and Sons. Ltd).

[10] Burger A M and Fiebig H-H 2014 Preclinical screening for new anticancer agents. In Rudek M A, Chau C H, Figg W, McLeod H L (Eds.) *Handbook anticancer pharmacokinetics and Pharmacodynamics* (New York: Springer) 23–38.

[11] Huang C-C, Lo C-P, Chiu C-Y, Shyur L-F 2010 Deoxyelephantopin, a novel multifunctional agent, suppresses mammary tumour growth and lung metastasis and doubles survival time in mice Br *J. Pharmacol.* **159** 856–71.

[12] Chang C L, Shen C C, Ni C L and Chen C C 2011 A new sesquiterpene from Elephantopus scaber *J. Hongkuang University* **65** 49–56.

[13] Kajstura M, Halicka H D, Pryjma J and Darzynkiewicz Z 2007 Discontinuous fragmentation of nuclear DNA during apoptosis revealed by discrete "sub-G1" peaks on DNA content histograms *Cytometry* Part A **71A** 125–31.

[14] Mattes M J 2007 Apoptosis assays with lymphoma cell lines: problems and pitfalls *Br. J. Cancer* **96** 928–36.

[15] Plesca D, Mazumder S and Almasan A 2008 DNA damage response and apoptosis *Methods Enzymol* **446** 107–22.

The immunomodulatory effect of *Zingiber cassumunar* ethanolic extract on phagocytic activity, nitrit oxide and reaxtive oxygen intermediate secretions of macrophage in mice

Nurkhasanah, R D Santoso, R Fauziah

Faculty of Pharmacy, Universitas Ahmad Dahlan, Yogyakarta, Indonesia

E-mail: nurkhas@gmail.com

Abstract. Immunomodulators could protect the body from a variety of infectious agents and boost immunity. *Zingiber cassumunar* rhizome or bangle potentially showed as an immunomodulator through increasing of macrophage activity *in vitro*. The objective of the study was to determine the effect of *Z. cassumunar* rhizome ethanolic extract on phagocytic activity, nitrite oxide (NO) and reactive oxygen intermediate (ROI) secretions in macrophages *in vivo*. A total of 200 g of *Z. cassumunar* rhizome was powdered, macerated in 96% ethanol and evaporated to get concentrated extract. Mice were divided into 5 groups as follow: the normal group was given by water only, the negative control group was given by a 0.94% CMC-Na suspension, the treatment groups were given by 250, 500 and 1000 mg/kgBW, respectively, of *Z. cassumunar* ethanolic extract. The extract was administered orally for 7 days. On the 8th day the mice were injected intraperitoneally 0.7 mg/kg BW of lipopolysaccharide. Four hours later macrophage was isolated. Furthermore, the determination of the phagocytic activity, NO and ROI secretions levels of macrophage were performed. The treatments of 250, 500 and 1000 mg/kg BW of *Z. cassumunar* ethanolic extract significantly increase the ROI and NO secretions levels ($p<0.05$), but did not increase the phagocytic activity ($p>0.05$) of macrophage. *Z. cassumunar* ethanolic extract have immunomodulatory effect in vivo.

Keywords: *Z.cassumunar*, imunomodulator, phagocytic activity, ROI, NO

1. Introduction

The environment around humans contains various types of pathogenic elements, such as bacteria, viruses, funguses, protozoa and parasites which can cause disease in human body. Infections that occur in normal people are generally short and rarely leave permanent damage. This is because the human body has a system called the immune system that responds and protects the body against the pathogenic elements. The immune response is depended on the ability of the immune system to recognize foreign molecules (antigens) which presented in pathogens and then generate appropriate reactions to exclude antigenic sources [1].

Increasing of immune response could be done by improving the function of the immune system using immunostimulant. Immunostimulants can increase the the body's resistance in fighting against various infections or to assist in the treatment of diseases associated with suppression of the immune system. Immunostimulants work by stimulating the main factors of the immune system, through

IOP Conf. Series: Materials Science and Engineering **259** (2017) 012007 doi:10.1088/1757-899X/259/1/012007

phagocytosis either through direct phagocytic mechanisms or by indirect mechanisms by releasing reactive oxygen compounds such as NO and ROI [2]. NO and ROI play a role in cooperating with macrophages destroy pathogens with dependent oxygen on the immune system.

Traditionally, *Z. cassumunar* has been used for treatment various deseases including antidiarrheal, fever, stomach, gout, asthma, and jaundice. Previous studies reported that *Z. cassumunar* Roxb have potential antioxidant activity [3–6]. *Z. cassumunar* also was reported to be an immunomodulator by increasing phagocytic activity in vitro [7,8]. The main components of the *Z. cassumunar* rhizome are essential oils [9] containing of chemical compounds such as sabinene, pinen, caryofillen oxide and karyofillen [10].

Understanding of mechanisms is an important part of developing extracts to herbal remedies which can be used in formal health services. *Z. cassumunar* has been used traditionally to treat various diseases but the mechanism of its action is still very limited. This study will examine the potential of ethanol extract of *Z. cassumunar* as an immunomodulator and assess the mechanism. This research is expected to provide benefits for the development and utilization of *Z. cassumunar* in health services.

2. Materials and Methods

2.1. Plant material

Z. cassumunar rhizome was purchased from local market of Pasar Beringharjo, Yogyakarta, Indonesia. The rhizome was macerated in 96% ethanol and evaporated to get a concentrated extract.

2.2. Animal treatment

The animal handling procedures in this study was approved by the Research Ethics Committee, Universitas Ahmad Dahlan, with approval number 011601011. The test animal used in this study was Swiss male mice, 8 weeks of age. The mice were divided into 5 groups: the normal group was given water only, the negative control group was given a 0.94% CMC-Na suspension only, the treatment groups were given with 250, 500 and 1000 mg/kg bw of *Z. cassumunar* ethanol extract. The extract was administered orally for 7 days. On the 8th day the mice were injected intraperitoneally of 0.7 mg/kgBW lipopolysaccharide, 4 hours later macrophages were isolated.

2.3. Macrophage isolation

The mice were fasted for 10-12 hours, then narcosed with chloroform. After that, mice was sprayed with disinfectant solution and placed in the supine position. The abdominal skin was opened and cleaned the peritoneum sheath with 70% alcohol. The 10 ml of RPMI medium was injected into the peritoneum cavity and waited for 3 minutes.

The peritoneal cavity fluid was aspirated with an injection syringe from a non-fat and distant part of the intestine. The aspirate was centrifuged at 1200 rpm, 4°C, for 10 min. The supernatant was removed, and the obtained macrophages resuspended by 1000 μl of complete medium. The number of cells was calculated by the hemocytometer with 10 μl of macrophages suspension.

The cell suspension was grown by 1000 μl into a 24-well microplate well with a density of $1x10^5$ cells/ml for NO secretion testing. For phagocytic activity test and ROI secretion, the cells were grown with density of $5x10^5$ cells/well into 6-well microplate. The cells was then incubated for 30 min in 5% CO_2 incubator with 37°C temperature, then 800 μl complete medium added into the susceptibility. Cell suspension for NO or ROI secretion test was incubated in 5% CO_2 incubator, 37°C for 24 hours.

2.4. Phagocytic assay

The medium in the microplate well was discard from cultured cell. Each wells was then added with 50 uL latex suspension and incubated for 60 min in 5% CO_2 incubator at 37°C. Following incubation, latex suspension was discard and cell was let in the room temperature for drying. Cell was then fixed using methanol for 3 min and methanol was discarded. After that, cells was stained by 10% Giemsa for 30 min. The cell was then washed with distilled water and observed using microscope with 400x magnification.

2.5. *Greiss assay for detection of NO secretion*
Greiss reagent was prepared by mixing Greiss A and Greiss B in the same amount. Greiss A was composed by of 0.1 g N-(1-naphthyl) ethylene diaminehydrochloride (NED) (Sigma N, 5889) in 100 ml of distilled water. Greiss B was prepared by mixing 1 g of sulfanilamide (Sigma N 5589) in 100 ml of 5% orthophosphoric acid. Greiss A and Greiss B must be stored in the dark place for protecting from direct light.

The culture suspension macrophage of 50 ul was placed onto each well in 96 well microplate. The sample was added by 50 ul greiss reagent, and let in room temperature for 15 min until the colour was changed. The absorbance of nitric oxide was observed on wavelength of 550 nm. Nitrite standard solution was used for calculating the concentration.

2.6. *ROI secretion assay*
The 50 µl of NBT solution was added to each wells. The cells was then added by 1 ml of PBS containing 125 ng/ ml of PMA in the middle of coverslip and incubated in 5% CO_2 incubator, 37°C for 60 min. The reagent was removed from the wells, and dried at room temperature. After cells were dry, cells were fixed with 1000 µl methanol for 30 seconds. After drying, cells were added by 200 µl neutral red spultion 2% for 15 minutes and dried at room temperature. After drying, cellswere rinsed with distilled water. The percentage of macrophage cells showing NBT reduction was calculated from about 100 cells examined with a 400 times magnification of light microscope.

3. Results and Discussion
The immune system is a mechanism by which the body maintains its integrity to protect against various pathogenic matter. The immune system consist of a specific and a nonspecific immune system. *Z. cassumunar* was reported to be a potential an immunomodulatory agent. One of chemical content was found in *Z. cassumunar* was curcumin. The dose treatment of extract in mice was equal to the treatmen of 18.08 mg/20g BW of curcumin . The dose can increase the NO secretion compared to negative control [11]. The treament of ethanol extract of *Z. cassumunar* for 14 days was also found to increase the phagocytic capacity of macrophages in mice induced by *Plasmodium berghei* [12].

In this research, lipopolysachcharide (LPS) was used as antigen for boosting the immune response. Injection of 1 mg/kg BW of LPS from *E. coli* O111B4 intraperitoneally could increase the C-reactive protein in mice serum as a part of acute inflammation reaction [13].

3.1. *Effect of Z. cassumunar rhizome extract on macrophage phagocytosis*
Phagocytic activity of macrophage is activated by the presence of antigens from macromolecules and pathogens. In this study the phagocytic activity of macrophage was observed by determining the macrophages which phagocytosed latex. The parameter used was Active Phagocytes Cells (SFA) refer to the number of macrophage cells that phagocytosed latex cells in 100 macrophage cells. The other parameter was the phagocytic capacity refer to the amount of latex which is phagocytosed in 100 cells macrophages and Phagocytosis Index (IF) refer to the average number of latex particles deposited in 100 macrophage cells.

Table 1 shows that the greater the dose of ethanol extract of *Z. cassumunar*, the greater the value of SFA, phagocytic capacity and phagocytosis index. This result indicate that the greatest activity of phagocytosis is present at a dose of 1000 mg/kgBW. This is probably due to the content of flavonoids and curcumin in the *Z. cassumunar* rhizome. Flavonoids potentially work against lymphokines produced by T cells which stimulate phagocyte cells to respond to phagocytosis [14].

The curcumin which is also found in the *Z. cassumunar* could enhance reactive oxygen species (ROS) which will further activate a signal involving peroxism proliferator activated receptor gamma (PPAR-γ) and NF-E2-related factor 2 (Nrf2). The activity of both signals on monocytes and macrophages results in increased expression of cluster defferensiation 36 (CD36) in order to increase macrophage phagocytosis. Activated macrophages receive signals from interferon alfa (INF-α) to produce inducible nitrite oxide synthase (iNOS). The enzyme catalyzes the conversion of L-arginine to L-citrulline which produces NO gas. Increasing of NO is associated with increasing of macrophage activity as phagocyte cells [11].

IOP Conf. Series: Materials Science and Engineering **259** (2017) 012007 doi:10.1088/1757-899X/259/1/012007

Phenylbutenoid which is found from the *Z. cassumunar* rhizome also show antioxidant activity through free radical scavenger and could increase phagocytic activity [7]. The highest value of SFA (%) among the groups treated by ethanol extract is 57,25% (in group which treated by 1000 mg/kgBW of extract). The previous study also reported that *Z. cassumunar* extract capable to increase the phagocytosis capacity in vitro study [15].

Tabel 1. The phagocytic activity of macrophage in mice treated with *Z. cassumunar* ethanol extract.

Group	SFA ±SD (%)	Phagocytic Capacity ±SD	Phagocytosis index ± SD
Normal	53.33±4.04	94.00±11.79	1.77±0.11
Negative control	52.00±3.00	92.33±4.62	1.77±0.04
Dose 250 mg/Kg bw	56.00±2.59	100.00±4.97	1.78±0.03
Dose 500 mg/Kg bw	56.25±3.59	100.50±14.20	1.78±0.14
Dose 1000 mg/Kg bw	57.25±3.86	110.75±19.65	1.92±0.22

The statistical analysis found that the number of active phagocyte cells (SFA), phagocytic capacity and phagocytosis index in this research did not show a significant difference ($p>0.05$).

3.2. The increasing of ROI secretion after treatment of Z. cassumunar extract

The measurement of peritoneal macrophage ability on secretion of ROI was measured by NBT Reduction assay containing PMA. PMA will stimulate the macrophages to secrete ROI. The presence of ROI (superoxide anion, $O2^-$) secretion exhibits the increasing of respiration and causes reduction of NBT to form a non-dissolved formazan precipitate in blue colour.

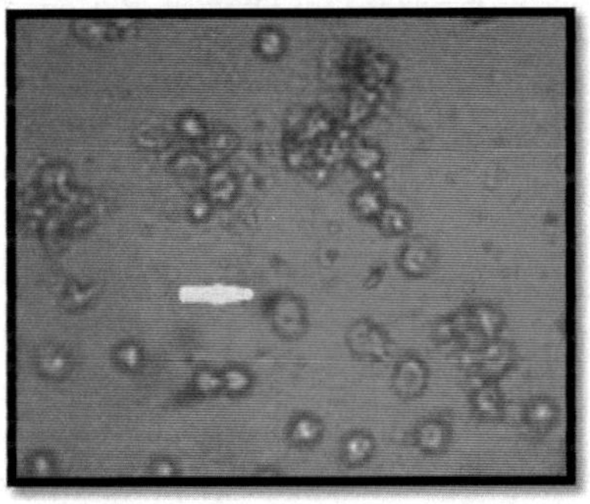

Figure 1. The microscopic feature of peritoneal macrophage with ROI secretion. ROI secretion showed by blue colour

Table 2. The percentage of ROI secretion of macrophage in mice treated with *Z. cassumunar* extract.

Groups	Percentage of ROI ± SD
Normal	19.00±2.00
Negative control	18.33±0.57
Dose 250 mg/kg bw	24.50±3.32^{a,b}
Dose 500 mg/kg bw	22.75±3.59
Dose1000 mg/kg bw	27.75±3.86^{a,b}

a : significant difference with normal group
b : significant difference with negative control group

The research found that treatment of LPS 0.7 mg/kg bw did not give any significant difference in ROI secretion compare to normal. The treatment of extract with dose of 250 and 1000 mg/kg bw increase the ROI secretion significantly compared to the negative control and normal. But statistical analysis of increasing dose in this study resulted no significant difference in ROI secretion.

3.3. The increasing of NO secretion after treatment of Z.cassumunar extract

Nitric oxide (NO) is an effective antibacterial effector in the immune system. The concentration of NO production secretion was found to increase after stimulating macrophages with IFN-γ and TNF-α. The IFN- γ is one of the cytokines which can induce NO secretion without stimulation by others, whereas TNF-α can not induce INOS without IFN- γ. NO is also used as an oxidative stress marker, which the accumulation of free radicals and inability of the antioxidant to eliminate the accumulation of free radicals cause imbalance production of reactive oxygen species [16].

NO is also recognized as an intercellular messenger that has been recognized as one of the most important players in the immune system. Cells of the innate immune system including macrophages, neutrophils and natural killer cells – use pattern recognition receptors to recognize the molecular patterns associated with pathogens. Activated macrophages then inhibit pathogen replication by releasing a variety of effector molecules, including NO. A large number of other immune-system cells produce and respond to NO. Thus, NO is important as a defense molecule against infectious microorganisms. It also regulates the functional activity, growth and death of many immune and inflammatory cell types including macrophages, T lymphocytes, antigen-presenting cells, mast cells, neutrophils and natural killer cells [17].

Previous study reported that the capability of curcumin, one of the active compound found in many Zingiber family, in increasing the reactive oxygen species (ROS) secretion by activating PPAR- *γ and Nrf-2* of macrophage and caused the increasing of CD36 [11].

The present study found that LPS could significantly increase the NO secretion compared to the control. LPS was recognized as antigen by macrophage and followed by stimulating the NO secretion. Treatment with *Z. cassumunar* extract decreased the NO secretion significantly (Table 3).

Tabel 3. The NO secretion level of macrophage in mice treated with *Z. cassumunar* extract

Group	NO secretion level Median (Minimum-Maximum)
Normal	3.914 (3.914-23.629)
Negative control	67.371 (34.514-111.09)[a]
Dose of 250 mg/kg bw	13.357(9.629-27.629)[b]
Dose of 500 mg/kg bw	30.371 (8.914-47.657)
Dose of 1000 mg/kg bw	7.657(3.229-87.514)

a : significant difference with normal group
b : significant difference with negative control group

The NO secretion levels of groups treated with 250, 500 and 1000 mg/ kg bw of extract was higher than that of the normal group, but lower than that of the group injected by 0.7 mg/kg bw of LPS. It means that the treatment of *Z. cassumunar* extract at all three doses decrease the levels of NO secretion of peritoneal macrophages peritoneum in mice induced by LPS 0.7 mg /kg bw. The previous study reported that intraperitoneal injection of LPS increased NO levels in serum macrophages [18]. The decrease of NO level after treatment with *Z. cassumunar* extract, could be resulted by the antioxidant compound of the extract. The curcumin compound reported to inhibit NO production in activated macrophages [19]. Curcumin also was reported to reduced the iNOS mRNA expression in LPS injected mice [20].

In vitro study confirmed the result, that *Z. cassumunar* could decrease the NO secretion in vitro in murine macrophage cell line RAW 264 [21]. Another active compound isolated from *Z. cassumunar Roxb* was also reported to inhibit NO production of peritoneal macrophage in LPS induce mice [22].

The high level of ROI and NO secretion indicate the good respon of immune system. Nevertheless excessive secretion of ROI and NO can result a reactive oxidant which involved the reactive oxygen compounds such as OH, OOH-, O_2, and H_2O_2. The presence of oxygen is important for the normal metabolism, but oxygen can increase the number of free radicals. Free radicals are highly reactive which can damage cells and can cause oxidative stress. Previous research reported the capability of cassumunin A and B which isolated from *Z. cassumunar* in prevented the hydrogen peroxide (H_2O_2)-induced decrease in cell viability caused by oxidative stress [23].

4. Conclusions

The administration of ethanol extract of *Z. cassumunar* rhizome for 7 days could increase ROI and decrease NO secretion levels but could not increase phagocytic activity ($p > 0.05$) of peritoneal macrophage in Swiss mice injected by LPS. These findings indicate that *Z. cassumunar* rhizome have immunomodulatory effect in vivo.

Acknowledgement

The authors would like to thank to Universitas Ahmad Dahlan for funding the research through the Fundamental Scheme in 2016.

References

[1] Kresno S 2001 *Imunologi: Diagnosis dan Prosedur Laboratorium* 4th ed. (Jakarta: Balai Penerbit Fakultas Kedokteran Universitas Indonesia).
[2] Baratawijaya 2001 *Imunologi Dasar* (Jakarta: Fakultas Kedokteran Universitas Indonesia).
[3] Bua-in S and Paisooksantivatana Y 2009 Essential oil and antioxidant activity of cassumunar ginger (Zingiberaceae: *Zingiber montanum* (Koenig) Link ex Dietr.) collected from various parts of Thailand *Kasetsart J. Nat. Sci.* **43** 467–75.
[4] Pal J, Pal S, Prakash O, Batra M and Kumar A 2011 Hepatoprotective and antioxidant activity of Zingiber chrysanthum Rosec rhizomes *Asian J. Tradit. Med.* **6**(6) 242–51.
[5] Vankar P S, Tiwari V, Singh L W and Swapana N. 2006 Antioxidant properties of some exclusive species of Zingiberaceae family of Manipur *EJEAF Che* **5**(2) 1318–22.
[6] Chanwitheesuk A, Teerawutgulrag A and Rakariyatham N 2005 Screening of antioxidant activity and antioxidant compounds of some edible plants of Thailand *Food Chem.* **92** 491–7.
[7] Chairul, Praptiwi and Chairul S M 2009 Phagocytosis effectivity test of phenylbutenoid compounds isolated from Bangle (*Zingiber cassumunar* Roxb.) rhizome *Biodiversitas* **10**(1) 40–3.
[8] Armiyant Y, Utami WS and Ameliana L 2014 Pengembangan ekstrak rimpag bangle (*Zingiber cassumunar* Roxb) terstandar menjadi granul effervesen sebagai terapi ajuvan untuk mencegah komplikasi pada malaria. Vol. 1 *Laporan penelitian* Fakultas Kedokteran Universitas Jember.
[9] Ayuningtyas D 2008 Aktivitas minyak atsiri rimpang bengle *Skripsi* Fakultas Kedokteran Universitas Diponegoro.
[10] Bhuiyan M N I, Chowdhury J U and Begum J. 2008 Volatile constituents of essential oils isolated from leaf and rhizome. *Bangladesh J. Pharmacol.* **3** 69–73.
[11] Nurmasari D P, Utami W S and Sulistyaningsih E 2014 Produksi Nitric oxide dan malondialdehyde pada mencit yang diinfeksi *Plasmodium berghei* (The role of bangle extract (Zingiber Cassumunar Roxb .) on nitric oxide and malondialdehyde of *Plasmodium beghei*-infected Mice) *e-Jurnal Pustaka Kesehat* **2**(3) 403–8.
[12] Rahmawati RM 2013 Uji imunostimulan ekstrak rimpang bangle kapasitas fagositosis makrofag dan produksi nitric oxide pada mencit yang diinfeksi *Plasmodium berghei Skripsi* Fakultas Kedokteran Universitas Jember.
[13] Adji D 2007 The effect of high lipid diet lipopolysaccharide and laparatomy surgery on the level of c-reactive protein serum in Sprague Dawley rats *J Sain Vet* **25**(2) 42-45.
[14] Nugroho YA 2012 The effect of *Piper betle* L, fruit (buah sirih), *Plectranthus scutellarioides* (L.) R. Br., leaf (daun miyana), honey and egg yolk on the improvement of macrophages activities and capacyties. *Media Litbang Kesehat* **22**(1):1–5.

[15] Chairul and Praptiwi. 2010 Uji efektivitas immunomodulator secara in-vitro melalui pengukuran aktivitas sel makrofag dan kapasitas fagositosis.

[16] Goodyear-bruch B C and Pierce J D, 2002 Oxidative stress in critical patients *Am. J. Crit. Care* **11**(6) 543–53.

[17] Tripathi P, Tripathi P, Kashyap L and Singh V 2017 The role of nitric oxide in inflammatory reactions. *FEMS Immunol. Med. Microbiol.* **51**(3) 443–52.

[18] Tunçtan B, Uludag O, Altug S and Abacioglu N 1998 Effects of nitric oxide synthase inhibition in lipopolysaccharide-induced sepsis in mice. *Pharmacol. Res.* **38**(5) 405–11.

[19] Brouet I and Ohshima H 1995 Curcumin, an anti-tumour promoter and anti-inflammatory agent, inhibits induction of nitric oxide synthase in activated macrophages *Biochem Biophys. Res. Commun.* **2** 533–40.

[20] Chan M, Huang H, M R F and Fong D 1998 In vivo inhibition of nitric oxide synthase gene expression by curcumin, a cancer preventive natural product with anti-inflammatory properties. *Biochem. Pharmacol.* **55**(12) 1955–62.

[21] Arafah E 2005 Perlindungan dan efek penyembuhan sediaan bangle (*Zingiber cassumunar Roxb.*) terhadap peradangan hati tikus serta mekanismenya pada sel makrofag dan limfosit, *Disertasi*, Sekolah Pascasarjana, Institut Pertanian Bogor.

[22] Nakamura S, Junko I, Matsuda H, Wakayama H, Pongpiriyadacha Y and Oshikawa M 2009 Structures of new phenylbutanoids and nitric oxide production inhibitors from the rhizomes of *Zingiber cassumunar Chem. Pharm. Bull.* **57**(11) 4–9.

[23] Nagano T, Oyama Y, Kajita N, Chikahisa L and Nakata M. 1997 New curcuminoids suffering from isolated oxidative using from zingiber cassumunar and protect study cells stress : A flow-cytometric rat thymocytes *Jpn. J. Pharmacol.* **75** 363–70.

Phytochemical screening and antioxidant activity of ethanolic extract and ethyl acetate fraction from basil leaf (*Ocimum basilicum* L.) by DPPH radical scavenging method

Warsi, A R Sholichah

Faculty of Pharmacy, Universitas Ahmad Dahlan, Yogyakarta, Indonesia

E-mail: warsisuryatmoko@gmail.com

Abstract. Basil leaf (*Ocimum basilicum* L.) contains various compounds such as flavonoid, alkaloid, phenol and essential oil, so it needs to be fractionated to find out the flavonoid compound with the greatest potential as an antioxidant. This research was aimed to know the chemical compound, antioxidant potential of ethanolic extract and ethyl acetate fraction from basil leaf. The basil leaf was extracted by maceration using ethanol 70 %. The crude extract was fractionated with ethyl acetate. The ethanolic extract and ethyl acetate fraction were screened of phytochemical content including identification of flavonoids, alkaloids and polyphenolics. The antioxidant activity of the ethanolic extract and ethyl acetate fraction were tested qualitatively with 2,2-diphenyl-1-picrylhydrazyl (DPPH) and phosphomolybdate. Its antioxidant activity was determined quantitatively using DPPH radical scavenging method. Phytochemical screening test showed that ethanolic extract and ethyl acetate fraction from basil leaf contain flavonoids, polyphenolics, and alkaloids. The qualitative analysis of antioxidant activity of ethanolic extract and ethyl acetate fraction from basil leaf showed an antioxidant activity. The IC_{50} value of ethanolic extract, ethyl acetate fraction and quercetin were 1,374.00±6.20; 389.00±1.00; 2.10±0.01µg/mL, respectively. The research showed that antioxidant activity of the ethyl acetate fraction more potential than the ethanol extract of the basil leaf, but less than quercetin.

Keywords: antioxidant, *Ocimum basilicum* L., basil leaf, ethanolic extract, ethyl acetate fraction, DPPH

1. Introduction

The life style changes lead to changes in dietary food. Fast food and instant food with preservative become favourite food. Preservatives and dyes contained in these foods can trigger the formation of free radicals which can cause various diseases. Free radicals are molecules with an unpaired electron. The presence of an unpaired electron, caused this molecule to become highly reactive [1]. Free radicals are also formed in natural processes in aerobic organism. Therefore, the body need antioxidant to neutralize free radicals. Antioxidant can stop the formation of free radicals and the chain reaction, which would result in cell damage. Antioxidant will transferred electrons or hydrogen to reduce its reactivity [2].

The body has a mechanism of defense against to free radicals, such as antioxidant enzymes (primary antioxidant), among are: superoxide dismutase (SOD), glutathione peroxidase (GPx) and glutathione reductase (GR). But if the amount of free radicals are excessive, the body can not against it. Thus the supply of antioxidants (secondary antioxidant) are needed to against these free radicals. The synthetic antioxidant such as butyl hydroxyl anisol (BHA) and butyl hydroxy toluene (BHT) are very effective and widely used for the processing industry, but have some side effects on human health, such as carcinogenesis promoters [3, 4].

Therefore, there has been much interest in the antioxidant of naturally occurring substances and develop it to replace synthetic antioxidant. Most plant derived foods have an antioxidant compounds that include vitamin C, vitamin E, polyphenols, carotenoids and ubiquinols [5]. One of the vegetables that has potential as a natural antioxidant is basil leaf (*O. basilicum* L. family Lamiaceae). It was reported that basil contains of flavonoids as naturally antioxidant [6]. The previous research reported on ethanolic extract of various kinds of vegetables including *O. basilicum* L., *Cichorium endivia* L.var.crispum Lam, *Amaranthus lividus* L.sensu Thell, and *Sauropus androgynus* L. Merr as antioxidants, and basil has the greatest an antioxidant potential [7]. The study of an antioxidant activity from water extract and ethanolic extract from *O. basilicum* L. which grown in Turkey were reported [8]. Another study was reported that basil extract contains phenolic compounds such as caffeic acid, dihydrokaempferol-3-O-glucoside, luteolin acetyl-glucuronide, dihydroxykaempferol-glycoside, caffeoyl ester, rosmarinic acid, carnosic acid, catechin, caffeoyl-3-O-rutiniside, rosmarinic acid, ferulic acid, apigenin and chlorogenic acid. These compounds potential as an antioxidant [9]. Basil also contains of essential oil which consist of eugenol, chavicol, linalool and a-terpineol. These compounds was reported as an antioxidant [10].

The purpose of this study was to identify of flavonoids, polyphenolics and alkaloids that contain in ethanolic extract and ethyl acetate fraction from basil leaf and evaluate its antioxidant activity. The antioxidant activity test in this study was performed using free radical scavenging method. The compound of 2,2-diphenyl-1-picrilhidrazil (DPPH) was used as free radical model.

2. Materials and Method

2.1. Chemicals

DPPH and quercetin were purchased from Sigma (Aldrich. Co, USA). All chemicals used were analitycal grade.

2.2. Plant materials, extraction and fractionation procedures

Basil leaf was obtained from local market at Muntilan Central Java, Indonesia. 1.7 kg wet basil leaf was dried in oven at 50°C until dry. For preliminary extraction, 200 grams powder of basil leaf was mixed with 800 mL petroleum ether (1:4) by magnetic stirred during 3 hours. The sample was left for 24 hours with periodic shaking. Then it was filtered over Whatman No.1 paper using Büchner funnel. The filtrate was removed, while the dregs was dried in the acidic wardrobe until dry and did not smell petroleum ether.

For ethanolic extraction, the residue was maceration using ethanol 70 % (1:4), then stirred by magnetic stirred during 30 minutes. The sample was left for 24 hours with periodic shaking. The extracts was filtered over Whatman No.1 paper using Büchner funnel. The residue was remacerated two times. The filtrate was collected and ethanol was removed by a Büchi rotary evaporator at 60°C. Then evaporation of extract was continued on waterbath at 60°C until became viscous.

For ethyl acetate fractionation, 5 g of viscous extract was inserted into a mortar with warm distilled water then stirred to form a suspension and transferred into a separating funnel. The extract suspension was added with ethyl acetate (1:1) and shaken to formed two layers. The ethyl acetate-soluble fraction that above was taken, then filtered over Whatman No.1 paper as filtrate I using Büchner funnel. While ethyl acetate unsolved fraction was refractionated with ethyl acetate two times until obtained filtrate II and III. The ethyl acetate soluble fraction were collected then evaporated by a Büchi rotary evaporator at 60°C. Then evaporation of fraction was continued on waterbath at 60°C until became viscous.

2.3. Phytochemical screening

2.3.1. Flavonoids identification

One drop of ethanolic extract and ethyl acetate fraction were dropped on filter paper and allowed to dry. The dry droplets of the samples on filter paper were steamed with ammonia. When the color change from dimly yellow to intense yellow showed that samples positive contain flavonoids.

Identification of flavonoids from ethanolic extract and ethyl acetate fraction using thin layer chromatography (TLC) were carried out on silica gel GF60 plates in a presaturated chamber. The solvent system was the mixture buthanol-acetic acid-water (BAW) (4: 1: 5). The identification of compounds using UV lamp at 254 and 366 nm. The silica gel GF60 plates was sprayed with $AlCl_3$, then observed under UV lamp at 254 and 366 nm.

2.3.2. Alkaloids identification

A volume of 1.0 mL ethanolic extract and ethyl acetate fraction were added 3 drops of Mayer's reagent. The positive samples with alkaloids contain would gave the white precipitate.

2.3.3. Polyphenolics identification

The ethanolic extract and ethyl acetate fraction as much as 1.0 mL were added 3 drops of $FeCl_3$ reagent. The samples positive contain polyphenols when a dark black colors appear.

2.4. Qualitative antioxidant activity

For this analysis, ethanolic extract and ethyl acetate fraction each 1.0 mL were added with 1.0 mL DPPH 0.15 mM reagent. The presence of antioxidant activity was characterized by the change of color from purple to yellow.

For the another analysis, ethanolic extract and ethyl acetate fraction each 1.0 mL were added 1.0 mL phosphomolybdate reagent. The presence of antioxidant activity was indicated by the discoloration from yellow to bluish green.

2.5. Quantitative antioxidant activity

2.5.1. Preparation of DPPH

The DPPH 1 mM solution was prepared by dissolving of 19.716 mg powder of DPPH with ethanol up to 50.0 mL in a volumetric flask. For DPPH 0.15 mM, 15.0 mL DPPH 1mM was added ethanol up to 100.0 mL.

2.5.2. Timing of operation determination

A volume of 1.0 mL samples were added 1.0 mL DPPH 0.15 mM in the test tube. The mixture were homogenized using a T 570 sonikator ultrasonic. The absorbance were measured using a Pharmaspec Shimadzu1800 UV-Vis spectrophotometer at 517 nm for 90 minutes.

2.5.3. Maximum absorption wavelength determination and negative control

The DPPH 0.15 mM as much as 1.0 mL was added with 1.0 mL of ethanol in the test tube. The mixture was homogenized using a T 570 sonikator ultrasonic and left to time of operation. The absorbance was measured using a Pharmaspec Shimadzu 1800 UV-Vis spectrophotometer at 450-600 nm. This data also was used as the absorbance of negative control.

2.5.4. Preparation of stock 0.2 mg/mL and series of quercetin

The quercetin 1 mg/mL was prepared by dissolving of 10.0 mg quercetin with ethanol up to 10.0 mL in a volumetric flask. For quercetin 0.2 mg/mL, 10.0 mL quercetin 1 mg/mL was added with ethanol up to 50.0 mL. The concentration series of quercetin were: 3, 2, 1.75, 1.5, 1.25 and 1 μg/mL.

2.5.5. Preparation of stock 2 mg/mL and series of ethanolic extract

The ethanolic extract 2 mg/mL was prepared by dissolving of 200.0 mg ethanolic extract with ethanol up to 100.0 mL in a volumetric flask. For series of ethanolic extract, each of 5.0, 4.375, 3.75, 3.125, 2.5 and 1.875 mL ethanolic extract 2 mg/mL were added ethanol up to 5.0 mL to obtain the solution with levels of 2, 1.75, 1.5, 1.25, 1 and 0.75 mg/mL, respectively.

2.5.6. Preparation of stock 1 mg/mL and series of ethyl acetate fraction

The ethyl acetate fraction 1 mg/mL was prepared by dissolving of 50.0 mg ethyl acetate fraction with ethanol up to 50.0 mL in a volumetric flask. For series of ethyl acetate fraction, each of 5.0, 2.5, 1.25, 0.5, 0.25 and 0.05 mL ethyl acetate fraction 1 mg/mL were added ethanol up to 5.0 mL to obtain the solution with level of 1, 0.5, 0.25, 0.1, 0.05 and 0.01 mg/mL, respectively.

2.5.7. The measurement of sample

The sample as much as 1.0 mL were added with 1.0 ml of DPPH 0.15 mM in the test tube. The mixture were homogenized using a T 570 sonicator ultrasonic and left to time of operation in the dark. The absorbance were measured using a Pharmaspec Shimadzu 1800 UV-Vis spectrophotometer at maximum wavelength against blank (2.0 mL ethanol). All tests were run in five times. The data were expressed as mean ± SD. Quercetin was used as a positive control.

2.6. Data analysis

The antioxidant capacity was expressed by % inhibition. The inhibition percentage of the DPPH radical was calculated according to the following equation:

$$\%\ \text{Inhibition DPPH} = \left(\frac{\text{Absorbance of control} - \text{Absorbance of sample}}{\text{Absorbance of control}}\right) \text{x } 100\ \%$$

The IC_{50} is defined as the concentration of the sample required to cause a 50 % decrease in initial DPPH radical absorbance. The IC_{50} was found after making the relationship between concentration and % inhibition. The IC_{50} value of ethanolic extract, ethyl acetate and quercetin were analysed statistically by SPSS 16 with 95% confidence level.

3. Results and Discussion

3.1. Extraction and fractionation

The active compounds of basil leaf were extracted by maceration method using ethanol 70% as solvent without heating process (Table 1). Therefore, it can prevent the occurrence destruction of thermolable active compounds. The deficiency of maseration method is the process of extraction takes a long time and saturation can occur. Stirring process can prevent any saturation. This extraction aims to attract the compounds that present in the simplicia.

Table 1. Data of drying, extraction and fractionation of basil leaf

Process	Initial Weight (gram)	Results (gram)	Yield (%)
Drying	1,700.00	342.21	20.13
Extraction	200.20	125.21	62.54
Fractionation	5.00	0.85	17.00

The advantage of using ethanol 70% is the compounds from less polar to polar can be withdrawn from simplicia, including flavonoids compounds [11]. The principle of extraction by maceration method is the diffusion from high to low concentration, the liquid of solvent can penetrate to the lining of the cell wall, then enter into the cell cavity that containing the active compounds. The active compounds in the cell were dissolved due to the difference in concentration between the solutions inside the cell (high concentration) with outside the cell (low concentration). The extract had characteristics viscous, brown color and typical smell as basil.

3.2. Phytochemical screening

3.2.1. Flavonoids identification

The results of flavonoids identification of ethanolic extract and ethyl acetate fraction with ammonia were the colors change from dimly yellow to intense yellow. It showed that samples contain flavonoids. Flavonoids were reacted with ammonia to form quinoid in this test. The yellow color occurs because of the formation of quinoid in β-ring that contain a longer conjugated double bond

[12]. The results of identification of flavonoids with ammonia are shown in Table 2. While the reaction of flavonoid with ammonia is shown in Figure 1.

Table 2. The results of flavonoids identification with ammonia

The samples	The samples solution	The samples with ammonia	Flavonoids content
Ethanolic Extract	 Dimly yellow	 Intense yellow	Positive
Ethyl acetate fraction	 Dimly yellow	 Intense yellow	Positive
Quercetin	 Dimly yellow	 Intense yellow	Positive

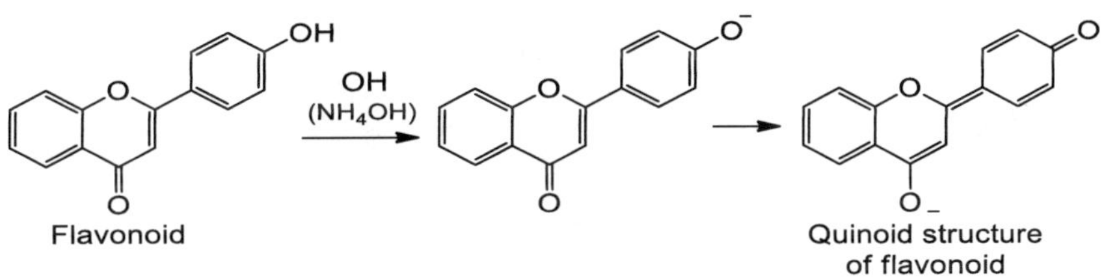

Figure 1. Reaction of flavonoid with ammonia [12]

The another test of flavonoids from ethanolic extract and ethyl acetate fraction were TLC method. The results of these tests obtained spots that were visually yellow after being sprayed using $AlCl_3$ reagents and read on 254 nm UV rays and 365 nm. Flavonoids were reacted with $AlCl_3$ to form falvonoid-Al^{3+} complex. The Rf of quercetin, ethanolic extract and ethyl acetate fraction were 0.81, 0.78 and 0.78, respectively. The Rf value are almost the same among the three samples. These results show that ethanolic extract and ethyl acetate fraction contains flavonoids. The identification of flavonoids with TLC are shown in Figure 2.

IOP Conf. Series: Materials Science and Engineering **259** (2017) 012008 doi:10.1088/1757-899X/259/1/012008

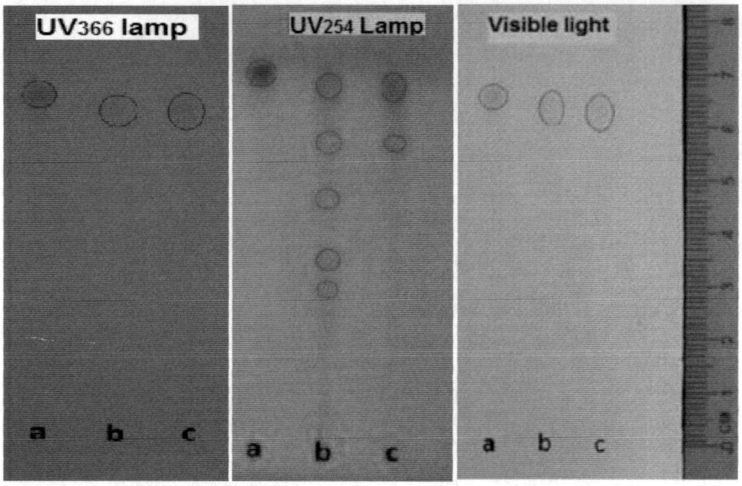

Figure 2. Test of flavonoids with TLC: quercetin (a), ethanolic extract (b), ethyl acetate fraction (c) on silica gel GF60 plates, BAW as mobile phase

3.2.2. Alkaloids identification

The results of alkaloids identification of ethanolic extract and ethyl acetate fraction with Mayer's reagent were appear white precipitate for all of the samples. The Mayer's reagents is a mixture of mercury with potassium iodide.

Table 3. The results of alkaloid identification with Mayer's reagent

The samples	The samples solution	The samples with Mayer's reagent	Alkaloids content
Ethanolic Extract	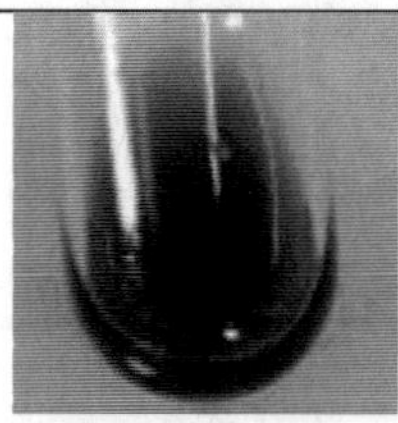 Brown-black	 White presipitate	Positive
Ethyl acetate fraction	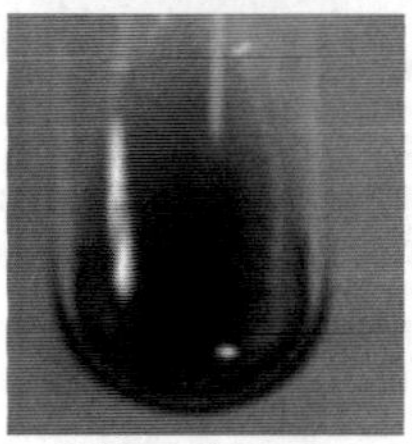 Green-brown	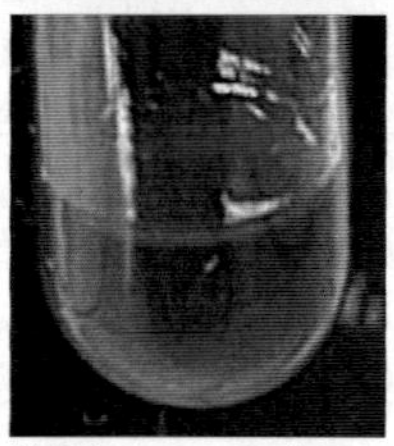 White presipitate	Positive

Alkaloids contain nitrogen atom with free electron pairs. This pair of free electrons reacts with the metal to form a covalent bond coordinate. The reaction of alkaloid test using this reagent may occur between the electrons of the nitrogen atom reacts with K^+ ion of dipotasium tetraIodomercurat so that forming a potassium-alkaloid complex which characterized by white presipitate [13]. The results of identification of alkaloids with Mayer's reagent are shown in Table 3. While the reaction of alkaloid with Mayer's reagent is shown in Figure 3.

IOP Conf. Series: Materials Science and Engineering **259** (2017) 012008 doi:10.1088/1757-899X/259/1/012008

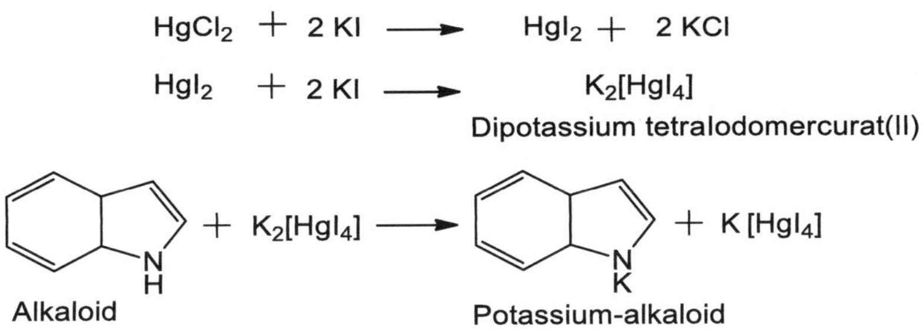

Figure 3. Reaction of alkaloid with Mayer's reagent [13]

3.2.3. Polyphenolics identification

The addition of $FeCl_3$ to the ethanolic extract and the ethyl acetate fraction that the colors change from brown-black and green brown respectively to black-brown. Its indicates that the samples contains of polyphenolics compounds. Polyphenolics compounds are containing OH groups. Therefore, it react with $FeCl_3$ to form flavonoid-Fe^{3+} complex [13]. The results of identification of polyphenolics with $FeCl_3$ are shown in Table 4. While the reaction of polyphenolics with $FeCl_3$ is shown in Figure 4.

Table 4. The results of polyphenolics identification with $FeCl_3$

The samples	The samples solution	The samples with $FeCl_3$	Polyphenolics content
Ethanolic extract	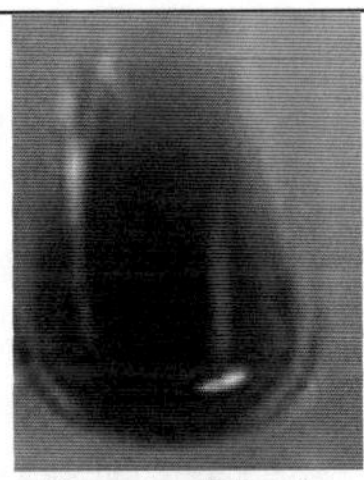 Brown-black	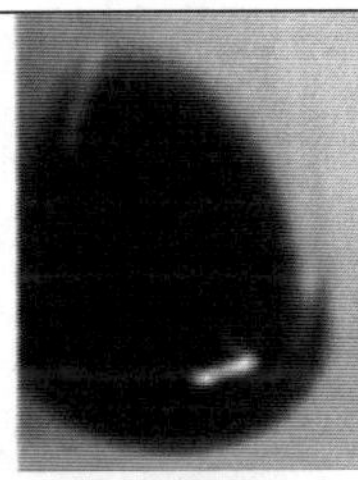 Black-brown	Positive
Ethyl acetate fraction	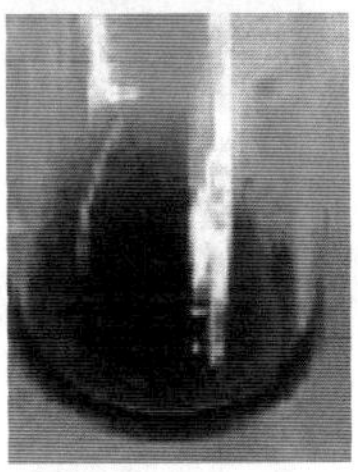 Green-brown	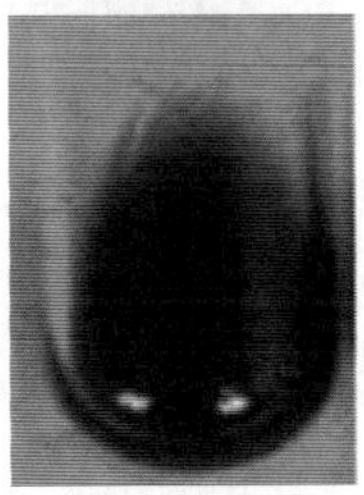 Black-brown	Positive

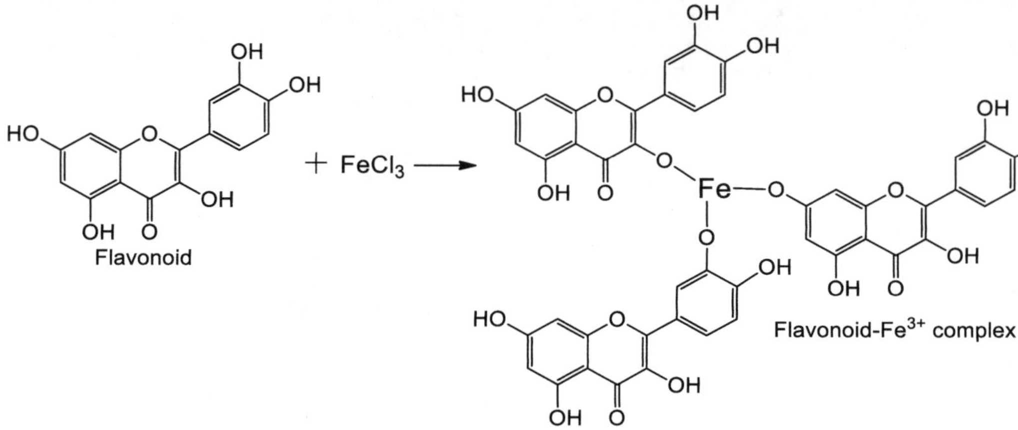

Figure 4. Reaction of polyphenolics with $FeCl_3$ [13]

3.3. Qualitative antioxidant activity

The qualitative antioxidant activity with DPPH were resulted the color change from purple to yellow-brown. It was showed that ethanolic extract and ethyl acetate fraction positive as an antioxidant. The results of qualitative antioxidant activity analysis with DPPH are shown in Table 5.

Table 5. The results of qualitative test antioxidant activity with DPPH

The samples	The samples solution	The samples with DPPH	Antioxidant activity
Ethanolic Extract	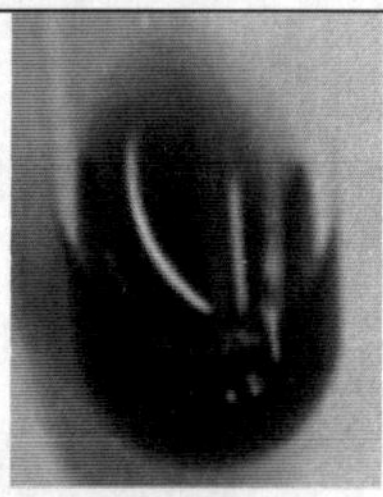 Purple	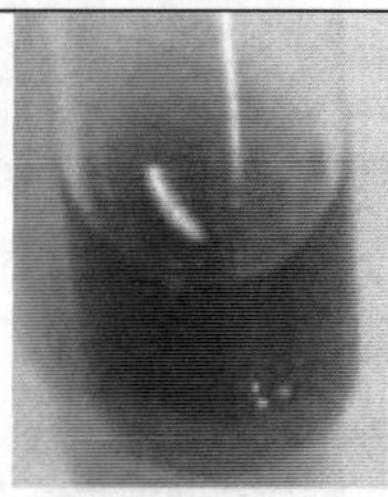 Yellow-Brown	Positive
Ethyl acetate fraction	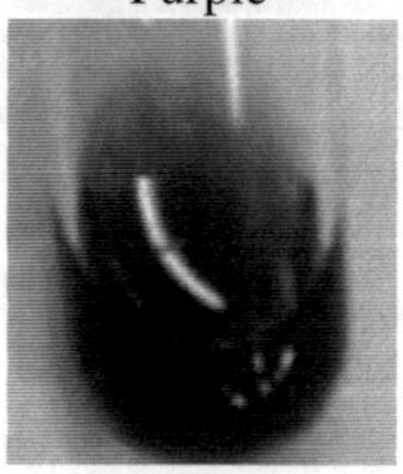 Purple	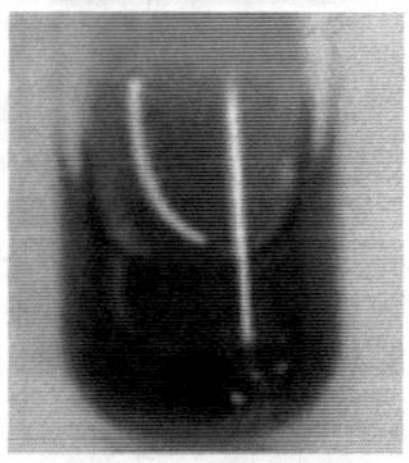 Yellow-Brown	Positive

The another test of an antioxidant activity is reductive to phosphomolybtae. The results of this test, both of the ethanolic extract and ethyl acetate fraction positive as anantioxidant. It was characterized by the change of color from yellow to green-blue. The reaction mechanism of this method was drawn by reaction of flavonoid with phosphomolybtae. Flavonoids which contain in the samples were oxidized by phosphomolybtae/ MO (VI) and formed MO (V). The oxidized flavonoids form react with H^+ in acidic condition (Figure 5). Antioxidant activity analysis qualitatively with phosphomolybdate are shown in Table 6.

Table 6. The results of qualitative antioxidant activity with phosphomolybdate

The samples	The samples solution	The samples with phosphomolybdate	Antioxidant activity
Ethanolic extract	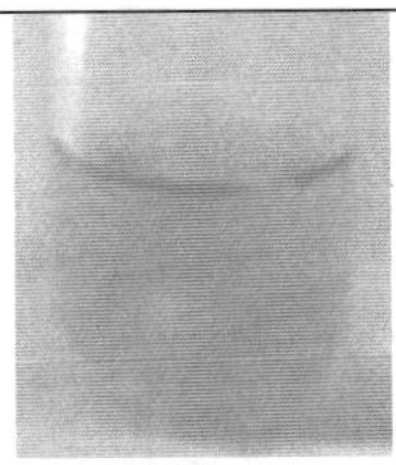 Yellow	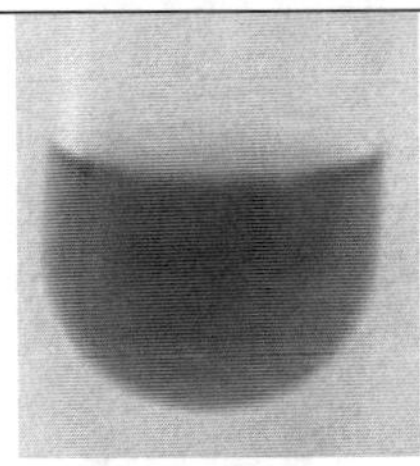 Green-Blue	Positive
Ethyl acetate fraction	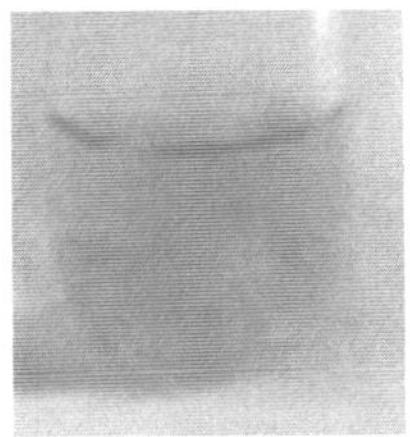 Yellow	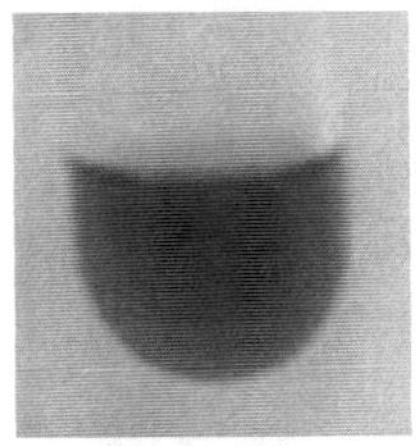 Green-Blue	Positive

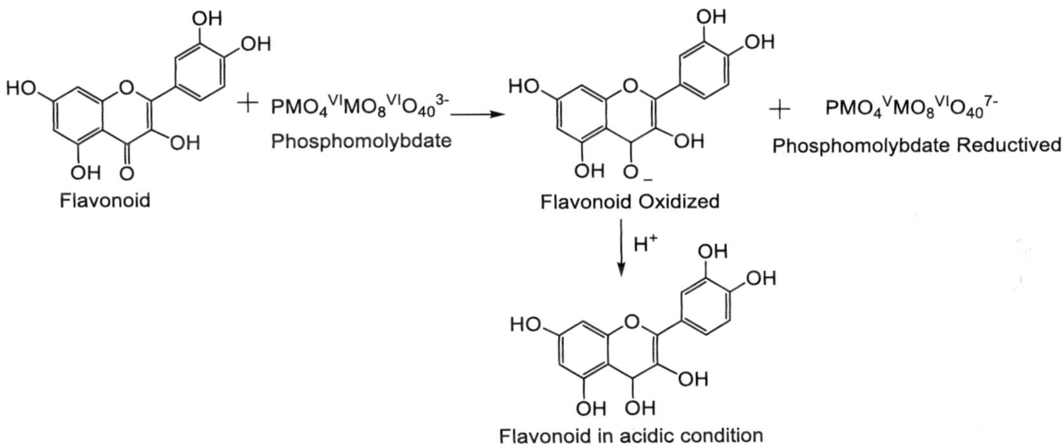

Figure 5. Reductive reaction of flavonoid to phosphomolybdate [14]

3.4. Quantitative antioxidant activity

The quantitative antioxidant activity analysis was performed using 2,2–diphenyl–1–pycrylhydrazyl (DPPH) radical scavenging method [15]. Quercetin was used as a standard because it has been shown have as free radical scavenging activity. The compound has phenolic (OH) groups at 3', 4', 3, 5, and 7 positions. The reaction mechanism of this method was illustrated by the flavonoid structure following the pattern of BHT reaction mechanisms with DPPH [15]. The reaction of type 1, flavonoid donate electron to DPPH and forming flavonoid radical species. Then flavonoid undergo resonance and quinoid structure are formed. While DPPH is forming $DPPH_2$ that stable and yellow color. The reaction of type 2, flavonoid after donating their electron to DPPH was formed a complex with DPPH (Figure 6).

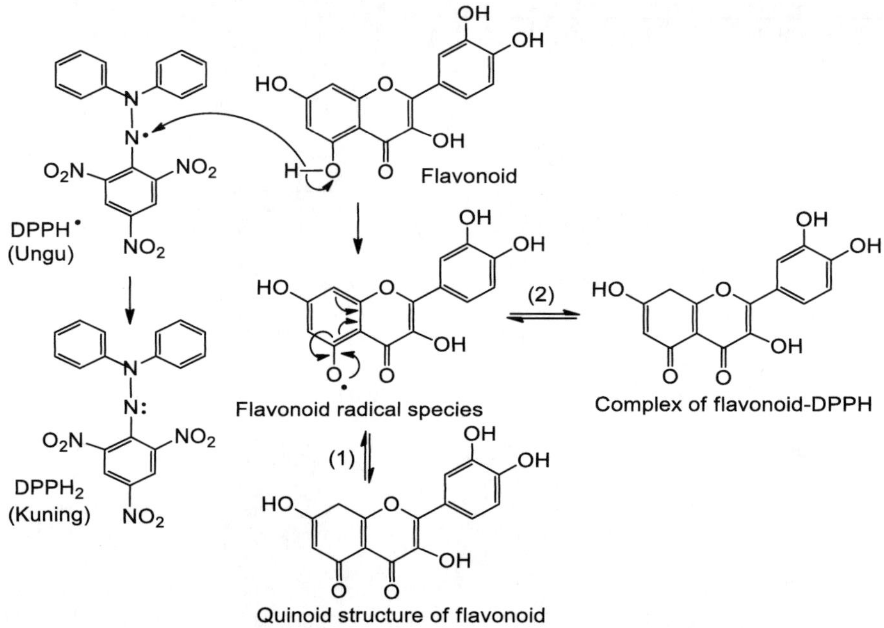

Figure 6. Reaction of flavonoid with DPPH [16]

The results of timing of operation determination from ethanolic extract, ethyl acetate fraction and quercetin were 72nd until 83nd, 76nd until 90nd and 68nd until 81nd minutes, respectively. The results of measurement of maximum absorption wavelength determination was obtained 516 nm. The quantitative antioxidant activity analysis with DPPH are shown in Table 7.

Table 7. The results of quantitative antioxidant activity analyzed with DPPH method

Sample	IC_{50} (μg/mL)	CV (%)
Ethanolic extract	1,374.00 ± 6.20	0.45
Ethyl acetate fraction	389.00 ± 1.00	0.26
Quercetin	2.10 ± 0.01	0.48

The results of IC_{50} value analysis statistically in normality test of ethanolic extract, ethyl acetate fraction and quercetin were obtained significance of 0.200>0.05; 0.200>0.05 and 0.072>0.05. It showed that data is normally distributed. The homogeneity test showed that data were homogeneous (p: 0.081>0.05). The results of analysis antioxidant activity showed that potential antioxidant activity of ethyl acetate fraction was higher than ethanolic extract of the basil leaf, but less than quercetin. The IC_{50} value of ethanolic extract, ethyl acetate fraction and quercetin were significantly different from each other (p: 0.000<0.05).

O. basilicum also active as antioxidant which performed with other methods, including ferric thiocyanate method, reductive capability by Fe^{3+}-Fe^{2+} transformation, ferrous ions chelating capacity and hydrogen peroxide scavenging activity [8].

4. Conclusions

Phytochemical screening test showed that ethanolic extract and ethyl acetate fraction from basil leaf contain flavonoids, polyphenolics, and alkaloids. The qualitative analysis of antioxidant activity of ethanolic extract and ethyl acetate fraction from basil leaf showed an antioxidant activity. The antioxidant activity of the ethyl acetate fraction more potential than the ethanolic extract of the basil leaf, but less than quercetin.

IOP Conf. Series: Materials Science and Engineering **259** (2017) 012008 doi:10.1088/1757-899X/259/1/012008

References

[1] Sarma A D, Mallick A R and Ghosh A K 2010 Free radicals and their role in different clinical conditions: an overview *IJPSR* **1** (3) 185–192.

[2] Wahlqvist M L 2013 Antioxidant relevance to human health mini review *Asia Pac. J. Clin. Nutr.* **22** (2) 171–176.

[3] Anagtostopoulou M A, Kefalas P, Papageorgiou V P, Assimepoulou A N and Boskou D 2006 Radical scavenging activity of various extracts and fractions of sweet orange peel (*Citrus sinensis) Food Chem.* **94** 19–25.

[4] Barlow S M 1990 Toxicological aspects of antioxidants used as food additives *In Food Antioxidants*; Hudson B J F, Ed (Amsterdam: Elsevier) 253–307.

[5] Shetty K and Wahlqvist M L 2004 A model for the role of the proline-linked pentose-phosphate pathway in phenolic phytochemical bio-synthesis and mechanism of action for human health and environmental applications *Asia Pac. J. Clin. Nutr.* **13** 1–24.

[6] Sarma D S K and Babu A V S 2011 Pharmacognostic and phytochemical studies of *Ocimum americanum J. Chem. Pharm. Res.* **3** (3) 337–347.

[7] Nurhaeni F 2012 Skrining aktivitas dan isolasi senyawa penangkap radikal 2,2-difenil-1-pikrilhidrazil dari kaun kenikir (*Cosmos caudatus*, H.B.K) *Thesis* Gadjah Mada University Yogyakarta Indonesia 114–118.

[8] Gülçin I, Elmastat M and Aboul-Enein H Y 2007 Determination of antioxidant and radical scavenging activity of basil (*Ocimum basilicum* L. family Lamiaceae) assayed by different methodologies *Phytother. Res.* **21** 354–361.

[9] Jayasinghe C, Gotoh N, Aoki T and Wada S 2003 Phenolics composition and antioxidant activity of sweet basil (*Ocimum basilicum* L.) *J. Agric. Food Chem.* **51** 4442–4449.

[10] Politeo O, Jukic M and Milos M 2007 Chemical composition and antioxidant capacity of free volatile aglycones from basil (*Ocimum basilicum* L.) compared with its essential oil *Food Chem.* **101** 379–385.

[11] Harborne J B 1996 *Phytochemical methods* 2nd print ITB Bandung 69–70.

[12] Robinson T 1995 *The Content of organic in high plant* print ITB Bandung.

[13] Simaremare E S 2014 Skrining fitokimia ekstrak etanol daun gatal (*Laportea decumma* (Roxb.) Wedd) *Pharmacy* **11** (01).

[14] Zengin G, Aktumsek A, Guler G O, Cakmak Y S and Yildistugay E 2010 Antioxidant properties of methanolic extract and fatty acid composition of *Centaurea urvillei* DC. subsp. Hayekiana Wagenitz *Rec. Nat. Prod.* **5** (2) 123–132.

[15] Sharma O P and Bhat TK 2009 Analytical methods, DPPH antioxidant assay revisied *Food Chem.* **113** 1202–1205.

[16] Bondet V, Brand–Williams W and Berset C 1997 Kinetics and mecanism of antioxidant activity using the DPPH free radical method *Lebens.-Wiss.u. Technol.* **30** 609–615.

IOP Conf. Series: Materials Science and Engineering **259** (2017) 012009 doi:10.1088/1757-899X/259/1/012009

Effect of subchronic administration of nutmeg (*Myristica fragrans* Houtt) ethanolic extract to hematological parameters in rat

M S Bachri[1], S Yuliani[1], A K Sari[2]

[1]Faculty of Pharmacy, Universitas Ahmad Dahlan, Yogyakarta, Indonesia

[2]ISFI Pharmacy Academy, Banjarmasin, South Kalimantan, Indonesia

E-mail: msaifulbachri@yahoo.co.id

Abstract. Nutmeg is dried kernel of broadly ovoid seed of *Myristica fragrans* Houtt. It has been mentioned in ethnomedical literature as aphrodisiac, stomachic, carminative, tonic, and nervous stimulant. In order to establish the safety of nutmeg, the effect of the repeated administration of nutmeg is needed. The study was aimed to determine the toxic effect of subchronic administration of nutmeg ethanolic extract to hematological parameters in rat. A total of 28 male adult Wistar rats divided into 4 groups. Group I as control was given by 0.5% CMC-suspension, group II, III, and IV were given by 50, 100, and 200 mg/kg bw, respectively, of nutmeg ethanolic extract. The treatments were administered daily for 31 days. On day 31 bloods were taken from orbital sinus. The hematological parameter consisted of the numbers of erythrocyte and leukocyte as well as hemoglobin and total protein levels were measured. The data were statistically analyzed by one way Anova followed by LSD test. All of observed hematological parameters in rats showed that there were no significant difference between the nutmeg ethanolic extract treated groups and control group. The result indicated that the subchronic administration of 50, 100, and 200 mg/kg bw of nutmeg ethanolic extract did not cause the change of hematological parameters in rat.

Keywords: *Myristica fragrans* Houtt, subchronic, hematological parameters

1. Introduction

Nutmeg is dried kernel of widely ovoid seed of *Myristica fragrans* Houtt. Nutmeg is an originally Indonesian plant come from the Banda and Maluku islands. Since ancient times in Indonesia, nutmeg has been used as a spice. Several medicinal and pharmacological properties of nutmeg have been reported. Nutmeg has aromatic, stimulant, narcotic, astringent, aphrodisiac, hypolipidaemic, anti-platelet aggregation, sedative and analgesic activities. Nutmeg seeds can also overcome sleep disorders, oxidative stress and be used as a remedy for stomach ache, rheumatism, and vomiting during pregnancy [1–3]. Eventhough nutmeg seed has been used for various symptons, nutmeg can cause intoxication. Nutmeg esential oil contains pine, sabincene, camphene, myristin, elemicin, isoelemicin, eugenol, isoeugenol, methoxyeugenol, safrole, diametric phenylpropanoids, lignas and

IOP Conf. Series: Materials Science and Engineering **259** (2017) 012009 doi:10.1088/1757-899X/259/1/012009

neolignas [4, 5]. Those compounds have been reported causing to paralysis, poisoning, drunkenness, anger, dizziness, falling, headaches, convulsions, hallucination and death [6].

Thus, the safety of nutmeg consumption has become important to be investigated. The toxic effects of the extract of nutmeg might occur with the administration of repeated doses in the animal system. The aim of this study was to investigate the toxic effect of sub chronic administration of nutmeg ethanol extract to the hematological parameters in male Wistar rat.

2. Methods

A total of 28 adult male Wistar rats (150-200 g), were randomly divided into 4 groups consisting of 7 rats per group. Rats were acclimatized for a week, and treated orally for 31 days. The treatment of each group as follows: Group I, as control, the rats were treated with a solution of 0.5% NaCl; Group II, the rats were treated with 50 mg/kg bw of nutmeg ethanolic extract; Group III, the rats were treated with 100 mg/kg bw of nutmeg ethanolic extract; Group IV, the rats were treated with 200 mg/kg bw of nutmeg ethanolic extract. At day31, the blood were taken from orbital sinus for hematological parameters measurement including the numbers of erythrocyte and leucocyte as well as the levels of hemoglobin and total protein.

2.1. The measurement of erythrocyte number

The number of erythrocyte were obtained as previous method [7]. The average number of erythrocytes was calculated by the formula: the number of cells counted x 10 (0.1 mm in) x 5 (⅕ from 1 mm^3) x 200 (1: 200). The number of erythrocyte were expressed as $10^6/mm^3$.

2.2. The measurement of leukocyte number

The number of leukocytes were evaluated according to the method described in the previous study [8]. The average number of leukocytes at day 31 in the all groups were calculated by the formula: the cells number counted x 20 (1:20) x 10 (0.1 mm in): 4 (number of squares in mm^2). The number of leukocytes were expressed as $10^3/mm^3$.

2.3. The measurement of Hb level

The level of Hb was evaluated according to the method described in the study by Guder, et al. [9] using spectrofotometer at 540 of wavelength.The formula for calculating hemoglobin (g/dL) is: the absorbance of standard/absorbance sample x standard concentration. The Hb level was expressed as g/dL.

2.4. The measurement of total protein level

The number of total protein was evaluated according to the method described in the study by Guder, et al. [9].The formula for calculating total protein (g/dL) was absorbance of sample/absorbance of standard x standard concentration. The total protein level was expressed as g/dL.

3. Result and Discussion

Observation of blood parameters are very important in the diagnosis of a functionals disorder of organs or tissues. Hematological examination may describe the function of body organs and physiological status. The data of the numbers of erythrocyte and leucocyte as well as the levels of hemoglobin and total protein in rat are presented in the Table 1.

Table 1. The number of erythrocyte and leukocyte as well as the level of hemoglobin and total protein (mean±SD) of blood in rat after administration of nutmeg ethanol extract for 31 days.

Groups	Dose (mg/kgBW)	Erythrocyte ($10^6/mm^3$)	Leukocyte ($10^3/mm^3$)	Hb (g/dL)	Total Protein (g/dL)
Control	-	7.848±0.409	7.86±0.801	10.640±0.551	8.252±0.595
Group 1	50	6.520±0.379*	7.70±1.134	14.910±4.134*	8.726±0.426
Group II	100	6.466±0.288*	7.47±1.082	13.914±0.867*	7.732±0.500
Group III	200	5.962±0.179*	7.20±1.259	14.124±1.034*	7.852±0.368

* = $p < 0.05$, significantly difference compared with control group

IOP Conf. Series: Materials Science and Engineering **259** (2017) 012009 doi:10.1088/1757-899X/259/1/012009

3.1. The number of erythrocyte

Erythrocytes are the red blood cells that travel in the blood. They are typically shaped as biconcave disks and manufactured in the bone marrow. The erythrocytes carry hemoglobin in the circulation. The main function of erythrocytes is to distribute oxygen to body tissues, and carry waste carbon dioxide back to the lungs. In humans, they survive in the circulation for an average of 120 days [10].

The result showed that there were significant difference between the rats treated extract groups and control group. The reduced of erythrocytes cells number may be caused by the suppression of blood cell synthesis or red cell lysis by saponins. This finding supportedthe prvious study that nutmeg contains saponin component which have toxic effects on the body's systems [11]. Saponin can cause hemolysis, the destruction of erythrocytes cell, resulting in the release of hemoglobin from the stromal erythrocytes (red blood grains). Hemolysis can be caused by a decrease in the surface tension of the cell membrane affected by several factors such as organic solvents, saponins, bile salts, soaps, enzymes, and other factors [12]. However, in the current study the number of erythrocytes in the extract treated groups showed still in the normal range (5.00-10,00 million/mm^3). Thus it can be concluded that administration of nutmeg seed ethanol extract of did not affect the number of erythrocytes.

The results of this study was different with the previous study [13] reported that administration of 100, 250 and 500 mg/kg bw of nutmeg ethanol extract for 14 days could significantly decrease ($p<0.05$) the number of erythrocytes. The difference of the study may be due to several factors such as the difference in concentration of ethanol solvent, the dose and duration of administration in rat.

3.2. The number of leukocytes

Leukocytes or white blood cells are the cells of the immune system involved in protecting the body against both infectious disease and foreign invaders. Immune system refers to the body's system ability to resist or eliminate abnormal cells or xenobiotic potentially damaging. Leukocytes and their derivatives can prevent pathogens invasion (disease-causing microorganisms, such as bacteria and viruses) through the process of phagocytosis, identifying and destroying cancerous cells that appear in the body and as cleansers that clean up body waste by phagocytically debris originating from dead cells or injury [14].

The result showed that there were no significant difference between the rats treated extract groups and control group. The result in all groups was within in normal range. The number of normal leukocytes of rat is 3.00-17.00 (10^3/mm^3) [15]. The administration of nutmeg ethanol extract for 31 days did not affect the number of leukocytes.

3.3. The level of haemoglobin (Hb)

Hemoglobin is a molecul which play an important role for storing and carrying the oxygen in the body. It contains a small amount of iron and heme protein. Hemoglobin has a strong appeal to oxygen. When red blood cells pass through the lungs, hemoglobin will bind oxygen from the air forming oxyhemoglobin [16].

The result showed that the mean of Hb level of group treated by 50 mg / kg bw of nutmeg ethanol extract was greater than the control group, however statistically there was no significant difference ($p>0.05$). Whereas the administration 100; 200 mg / kg bw of extract and the dose of 200 mg / kg bw significantly increase the hemoglobin levels compared to the control group ($p<0.05$). However, the increase in hemoglobin levels are still within the normal range (11.0-19.0 g/dL) [15]. Thus it can be concluded that administration ethanol extract of nutmeg seed did not change on blood hemoglobin level.

3.4. The total protein level

Plasma total proteins are a group of plasma constituents that are not simply transported. These essential components in normal state remain in the plasma, where they perform many useful functions. Because of a large plasma constituent, plasma proteins usually did not exit to the pores in capillary walls. Unlike other plasma constituents that water soluble, plasma proteins are in the form of colloidal dispersions [17].

The decreased of total protein (called hypoproteinemia) might be caused by prolonged malnutrition, starvation, low protein diet, malabsorption syndrome, gastrointestinal tract cancers, ulcerative colitis, Hodgkin's disease, severe liver disease, severe burns and water intoxication. Meanwhile the increased of total protein levels (hyperproteinemia) can be caused by dehydration (hemoconcentration), multiple myeloma, vomiting, diarrhea, respiratory distress syndrome, and sarcoidosis. Serum protein levels may return to normal if the client is adequately hydrated

The result showed that there were no significant difference between the rats treated extract groups and control group in total protein. The result in all groups were within in normal range. The normal total plasma protein level of rat was 5.9 – 8.4 g/dL [15]. Thr administration of ethanol extract of nutmeg seed at all doses did not affected protein total level.

4. Conclusion

The sub chronic administration of 50, 100 and 200 mg/kg bw of nutmeg ethanol extract for 31 days did not cause the change of the numbers of erythrocytes and leucocytes as well as the levels of hemoglobin and total protein in blood of rat.

Acknowledgement
The authors would like to acknowledge the Directorate of Higher Education, Ministry of Education and Culture, Indonesia for partially funding the research through Fundamental Grant Scheme (No. 522.5/K5/KL/2012).

References

[1] Tajuddin, Ahmad S, Latif A, Qasmi I A and Amin K M Y 2005 An experimental study of sexual function improving effect of *Myristica fragrans* Houtt. (nutmeg) *BMC Complement. Altern. Med.* **5** 16.

[2] Akinboro A, Mohamed K B, Asmawi M Z, Sulaiman S F and Sofiman O A 2011 Antioxidants in aqueous extract of *Myristica fragrans* (Houtt.) suppress mitosis and cyclophosphamide-induced chromosomal aberrations in *Allium cepa* L. cells *J. Zhejiang Univ. Sci.* B **12** 915–22.

[3] Arulmozhi D K, Kurian R, Veeranjaneyulu A and Bodhankar S L 2007 Antidiabetic and antihyperlipidemic effects of *Myristica fragrans*. in animal models *Pharm. Biol.* **45** 64–8.

[4] Isogai A, Murakoshi S, Suzuki A and Tamura S 1973 Structures of new dimeric phenylpropanoids from *Myristica fragrans* Houtt. *Agric. Biol. Chem.* **37** 1479–86.

[5] Janssens J, Laekeman G M, Pieters L A, Totte J, Herman A G and Vlietinck A J 1990 Nutmeg oil: identification and quantitation of its most active constituents as inhibitors of platelet aggregation *J. Ethnopharmacol.* **29** 179–188.

[6] Stein U, Greyer H, and Hentschel H 2001 Nutmeg (myristicin) poisoning—report on a fatal case and a series of cases recorded by a poison information centre *Forensic Sci. Int.* **118** 87–90.

[7] Dharmawan N S 2002 *Pengantar Patologi Klinik Veteriner, Hematologi Klinik.* Univ. Udayana Denpasar.

[8] M N, G K, P P, A HK, and C VN 2008 Allethrin-induced biochemical changes and properties of human erythrocyte membrane *Afr. J. Biochem. Res.* **2** 024–9.

[9] Guder WG, Narayanan S, Wisser H and Zawta B 2008 *Samples: from the patient to the laboratory: the impact of preanalytical variables on the quality of laboratory results.* John Wiley & Sons.

[10] Snyder G K and Sheafor B A. 1999 Red blood cells: centerpiece in the evolution of the vertebrate circulatory system. *Am. Zool.* **39** 189–198.

[11] Effraim K D, Salami H A and Osewa T S 2000 The effects of aqueous leaf extract of *Ocimum gratissium* on haematological and biochemical parameters in rabbits *Afr. J Biomed. Res.* **3** 175–179.

[12] Gauthier C, Legault J, Girard-Lalancette K, Mshvildadze V and Pichette A 2009 Haemolytic activity, cytotoxicity and membrane cell permeabilization of semi-synthetic and natural lupane- and oleanane-type saponins *Bioorg. Med. Chem.* **17** 2002–8.

[13] Bamidele O, Akinnuga A M, Alagbonsi I A, Ojo O A, Olorunfemi J O and Akuyoma M A 2011 Effects of ethanolic extract of *Myristica fragrans* Houtt. (nutmeg) on some heamatological indices in albino rats. *Int. J. Med. Med. Sci.* **3** 215–218.

[14] Walters M D S, Matthei I U, Kay R, Dillon M J and Barratt T M 1989 The polymorphonuclear leucocyte count in childhood haemolytic uraemic syndrome *Pediatr. Nephrol.* **3** 130–4.

[15] Sharp P and Villano J 2012 *The Laboratory Rat Second Edition.* CRC Press Taylor and Francis Group.

[16] Bunn H F, May M H, Kocholaty WF and Shields CE. 1969 Hemoglobin function in stored blood *J. Clin. Invest.* **48** 311–21.

[17] Putnam F 2012 The Plasma Proteins V3: *Structure, Function, and Genetic Control.* Elsevier.

Determination of total phenolic content and antioxidant activitity of methanol extract of *Maranta arundinacea* L fresh leaf and tuber

A Kusbandari, H Susanti

Faculty of Pharmacy, Universitas Ahmad Dahlan, Yogyakarta, Indonesia

E-mail: kusbandari80@yahoo.com

Abstract. *Maranta arundinacea* L is one of herbaceous plants in Indonesia which have flavonoid content. Flavonoids has antioxidants activity by inhibition of free radical oxidation reactions. The study aims were to determination total phenolic content and antioxidant activity of methanol extract of fresh leaf and tuber of *M. arundinacea* L by UV-Vis spectrophotometer. The methanol extracts were obtained with maceration and remaseration method of fresh leaves and tubers. The total phenolic content was assayed with visible spectrophotometric using Folin Ciocalteau reagent. The antioxidant activity was assayed with 1,1-diphenyl-2-picrilhidrazil (DPPH) compared to gallic acid. The results showed that methanol extract of tuber and fresh leaf of *M. arundinacea* L contained phenolic compound with total phenolic content (TPC) in fresh tuber of 3.881±0.064 (% GAE) and fresh leaf is 6.518±0.163 (% b/b GAE). IC_{50} value from fresh tuber is 1.780±0.0005 μg/mL and IC_{50} fresh leaf values of 0.274±0.0004 μg/mL while the standard gallic acid is IC_{50} of 0.640±0.0002 μg/mL.

Keywords: *M. arundinacea* L, total phenolic, antioxidant activity.

1. Introduction

The modernization era brings the impact of changes in people's lifestyle. Protection against disease can be done with the provision of antioxidants [1]. Antioxidants are compounds that can inhibit reactive oxygen species/ reactive nitrogen species (ROS/ RNS) as well as free radicals. Moreover, antioxidants can prevent diseases associated with free radicals such as carcinogenesis, cardiovascular diseases and aging [2].

Antioxidant mechanism could neutralize and destroy the compounds in the oxidation process [3]. The sources of antioxidant can be either synthetic or natural[4]. The antioxidant compounds isolated from plants are generally phenolic compounds or polyphenols in the form of flavonoids, cinnamic acid derivatives, coumarin, and tocopherol. The natural antioxidant compounds of polyphenols can be reducers, free radical encapsulation, metal chugs and dampers of singlet oxygen [5].

M. arundinacea L included in the family Maranthaceae [6]. *M. arundinacea* L bulbs used to make starch to treat ulcer peptic [7] . It is also reported to treat the diarrhea, pain and antioxidant [8]. The tuber has an effect as anti-cholesterol [9] and has anti-ulcer effects [7]. And they has 32 glycemic index (IG) which belongs to food category with low glycemic index [10]. The *M. arundinacea* L was originated from the South American region of West Brazil. The tuber contains high carbohydrates

IOP Conf. Series: Materials Science and Engineering **259** (2017) 012010 doi:10.1088/1757-899X/259/1/012010

which can be utilized as thickener, stabilizer, food ingredient, condiments, soup, candy, pudding and ice cream [11]. *M. arundinacea* has phenolic, flavonoid, alkaloids and saponin compound that are potential as antioxidants [12].

Research on flavonoid levels present in the ethanol extract of dried tubers, leaves and flowers have a total value of 390±11mg GAE/100g phenol, total flavonoids 290± mg QE / 100g and total flavonol 150 ± 9 mg QE / 100g [13]. In addition, the ethanol extract which obtained by percolation from the mixture of rhizomes, leaves, and flowers of *M. arundinacea* L from India produces the IC_{50} value of 293.4 µg / mL. In addition, antioxidant activity with ABTS, hydrogen peroxide and radical oxide nitrate with IC_{50} values was 297.4, 336.1, and 258.7 µg / mL [14]. The methanol extract of tuber contains total phenol and flavonoid total of 0.16% and 0.15% [15].

The present study used the fresh leaves and tubers without drying and pollination to get the greater content of phenolic compound.

2. Materials and Methods

2.1. Plant material

The study was carried out on fresh tubers and leaves of *M. arundinacea* L . It was found from Banguntapan, Bantul district of Yogyakarta Indonesia on May 2016. The plants was washed and chopped and then made extract without driying.

2.2. Preparation of extracts

A total of 250.0 grams of fresh samples (tuber and leaf) which have been sliced were macerated and remaserated with methanol and then evaporated to obtain a concentrated extract.

2.3. Determination of total phenolic content

Determination of total phenol content in tuber extract and fresh leaf was done by using Folin Ciocalteu. A total of 25 mg of methanol extract was dissolved in mixture of methanol : distilled water (1: 1). The solution was then taken 300 µL and added 1.5 mL of the folin reagent (1:10), then shaken out and allowed to stand for 3 minutes. The solution was added with 1.2 mL of a 7.5% Na_2CO_3 solution was shaken out homogeneously and at room temperature. The absorbance was measured at a wavelength of 730 nm using a spectrophotometer. The gallic acid was used as standard. The total phenolic content is expressed with % equivalent of gallic acid (% w/w GAE).

2.4. DPPH radical scavenging activity

Free radical activity with DPPH was done by mixing 1 mL sample and solution of DPPH 0.15mM 1 mL then homogenized and let stand in dark place and measured its absorbance with a spectrophotometer at 516,0 nm. Gallic acids were used as a comparison.

3. Results and Discussion

3.1. Plant extraction

The extract was obtained through the process of maceration and remaseration by using methanol. In this study, methanol was used for isolating the secondary metabolites in the simplicia[15]. Methanol is a universal solvent, so it can dissolve in the polar and nonpolar compounds. Methanol can isolate the alkaloids, steroids, saponins, and flavonoids from plants [17]. The rendemen of extracts was presented in Table 1.

Table 1. The rendemen of extracts of fresh *M. arundinacea L* leaf and tuber

part of plant	Weight material (gram)	Weight extract	Rendemen (%)
leaves	1500	109.5	7.3
Tuber	600	80.35	13.39

3.2. Determination of total phenolic content

The total phenolic content is expressed by % gallic acid (GAE). Total phenolic determinations were performed using a visible spectrophotometer [18]. The results showed that total phenol content was presented in Table 2.

Table 2. Total phenolic content of fresh tuber and fresh leaves of *M. arundinacea* L

replication	fresh tuber extract (g GAE/100g extract)	fresh leaf extract (g GAE/100g)
1	3.875	6.543
2	3.942	6.683
3	3.948	6.660
4	3.841	6.391
5	3.798	6.313
Average	**3.881**	**6.518**
SD	0.064	0.163
CV	1.667	2.501

Total phenol in the leaf is higher than tuber because the flavonoid is more content. While the carbohydrate content in the tuber is higher. The total phenolic content and total flavonoid content of the *M. arundinacea* rhizome extract were found to be 0.16g/100 g and 0.15g/100g respectively [15].

3.3. *DPPH radical scaveging activity*

The parameter of antioxidant activity was used to IC_{50}. An antioxidant activity describes how large compounds can capture free radicals in the body. The antioxidant activity of fresh leaf extract and fresh tuber extract was presented in Table 3.

Table 3. The antioxidant activity of *M. arundinacea* L leaf extract and tuber extract.

Name	value IC_{50} (µg/mL) ± SD
Standard gallic acid	0.640±0.0002
Fresh tuber extract	1.780±0.0005
Fresh leaf extract	0.274±0.0004

IC_{50} values of *M. arundinacea* extract and standard ascorbic acid for DPPH, scavenging activity were found to be 30.28 µg/mL and 18.26µg/mL respectively [15]. The ethanolic extract of *M. arundinacea* L was screened for antioxidant activity using 1,1-diphenyl-2-picryl hydroxyl (DPPH) scavenging assay with IC_{50} value of 293.4 µg/mL respectively [15].

4. Conclusions

The methanol extract from *M. arundinacea* L of fresh leaves and the fresh tuber have antioxidant activity with DPPH methods.

Acknowledgment

The authors thank to Research and Development Institute of Universitas Ahmad Dahlan for funding this research.

References

[1] Rohman A and Riyanto S 2005 Daya antioksidan ektrak etanol daun kemuning (*Murraya paniculata* (L) Jack secara invitro *Majalah Farmasi Indonesia* **16(3)** 136-140.

[2] Halliwell B and Gutteridge J M C 2000 *Free Radical in Biology and Medicine* (New York: Oxford University Press).
[3] Suryanto E 2012 *Fitokimia Antioksidan* (Surabaya: Putra Media Nusantara).
[4] Gordon I 1994 *Functional Food Food Design Pharmafood* (New York: Champman dan Hall).
[5] Kumalaningsih 2007 *Antioksidan dan Penangkal Radikal Bebas* (Jakarta: Trubus Agrisarana).
[6] Backer A and Brink R C B V D 1965 *Flora of Java* Vol 3 NVP (Netherlands: Noordhoof Groningen)
[7] Astuti RW 2008 Uji Efek Antiulcer Perasan Umbi Garut (Maranta arundinaceae L) pada Tikus Putih Jantan Galur Wistar *Skripsi* Fakultas Farmasi Universitas Muhammadiyah Surakarta Surakarta.
[8] Kumalasari I D, Harmayani E, Lestari L A, Raharjo S, Asmara W and Nishi K 2012 Evaluation of immunostimulatory effect of the arrowroot (*Maranta arundinacea*. L) in vitro and in vivo *Cytotechnology* **64**(2) 131-7.
[9] Aswan BS 2008 Daya hipokolesterolemik tepung umbi garut (*Marantha arundiceae*) secara In Vivo *Skripsi*, STP Fakultas Teknologi Pertanian Institut Pertanian Bogor Bogor.
[10] Utami 2008 Kajian indeks glisemik dan kapasitas in vitro pengikatan kolesterol dari umbi suweg (*Amorphophallus companulatus* BI) dan umbi garut (*Maranta arundinaceae* L) *Skripsi* Institut Pertanian Bogor, Bogor.
[11] Sediaoetama A D 1989 *Ilmu Gizi* (Jakarta: Dian Rakyat).
[12] Ramadhani M R, Bachri M S and Widyaningsih W 2017 Effects of ethanolic extract of arrowroot tubers *Maranta arundinacea* L on the level of MDA, SGPT and SGOT in ethanol induced rats JKKI **8**(1) 10-18.
[13] Nishaa S, Visnupriya M and Gopalakrishnan V K 2013 Qualtative assesment of ethanolic extract of *Maranta arundinacea* L Rhizome using HPTLC *IRJP* **4**(2) 76-83.
[14] Nishaa S, Vishnupriya M, Sasikumar JM, Christabel HP and GopalakrishnanVK 2012 Antioxidant activity of ethanolic extract of *Maranta Arundinacea* L tuberous rhizome *Asian J. Pharm. Clin. Re.s* **5**(4) 85-88.
[15] Ruba A A and Mohan V R 2013 Evaluation of total phenolic and flavonoid contents and in vitro antioxidant activity of rhizome of *Maranta arundinaceae L J. Pharmasm.* (4)2 p 3914-3928.
[16] Anonim 2008 *Farmakope 136-140 Herbal Indonesia Edisi 1* (Departemen Kesehatan Republik Indonesia Jakarta).
[17] Thompson E B 1985 Drug Bioscreening *Graceway Publishing Company* Inc America 40: 118.
[18] Chun O K, Kim Do and Lee C Y 2003Superoxide radicals scavenging activity of the mayor polyphenoles in fresh Plums *J. Agric. Food Chem.* (51) 5032-9.

HPLC determination of caffeine in coffee beverage

B E P Fajara, H Susanti

Faculty of Pharmacy, Universitas Ahmad Dahlan, Yogyakarta, Indonesia

E-mail: susantihari@gmail.com

Abstract. Coffee is the second largest beverage which is consumed by people in the world, besides the water. One of the compounds which contained in coffee is caffeine. Caffeine has the pharmacological effect such as stimulating the central nervous system. The purpose of this study is to determine the level of caffeine in coffee beverages with HPLC method. Three branded coffee beverages which include in 3 of Top Brand Index 2016 Phase 2 were used as samples. Qualitative analysis was performed by Parry method, Dragendorff reagent, and comparing the retention time between sample and caffeine standard. Quantitative analysis was done by HPLC method with methanol-water (95:5v/v) as mobile phase and ODS as stationary phasewith flow rate 1 mL/min and UV 272 nm as the detector. The level of caffeine data was statistically analyzed using Anova at 95% confidence level. The Qualitative analysis showed that the three samples contained caffeine. The average of caffeine level in coffee bottles of X, Y, and Z were 138.048 mg/bottle, 109.699 mg/bottle, and 147.669 mg/bottle, respectively. The caffeine content of the three coffee beverage samples are statistically different ($p<0.05$). The levels of caffeine contained in X, Y, and Z coffee beverage samples were not meet the requirements set by the Indonesian Standard Agency of 50 mg/serving.

Keywords: caffeine, beverage, coffee, HPLC

1. Introduction

Caffeine is an alkaloid compound of xanthine (Purine bases) which is naturally present in coffee. Caffeine has pharmacological effects, such as stimulating the central nervous system, eliminating drowsiness, and increasing concentration. However, excessive usage of caffeine can cause heartbeat, stomach upset, trembling hands, restlessness, diminished memory, and difficulty sleeping [1]. And if consumed above 500 mg (in one drink), will lead to poisoning [2]. And during the years 1995 to 2008, as many as 52.515 peoples in the world died from excessive coffee consumption [3].

To ensure the safety of coffee which is soled in the market, the National Standardization Agency (BSN) has set standards for caffeine levels in coffee ranging from 0.45-2% w/w (SNI 01-3542-2004) or 150 mg/day and 50 mg/serving (SNI 01-7152-2006) [4]. While based on the FDA (Food Drug Administration) [5], the dose of caffeine for adult is allowed <400 mg/day. The coffee which contained high levels of caffeine need to be decaffeinated, to suppress the activity of caffeine in the body [6].

Quantitative analysis of caffeine levels in some beverages and non-beverage products has been widely practiced by previous researchers with various methods, such as the determination of caffeine levels in cola soft drinks by HPLC, and the determination of caffeine levels in black coffee by the UV-Visible spectrophotometric method [7]. The good analytical method is required to provide accuracy, accuracy, selectivity, and high speed. HPLC becomes the preferred method because it meets these specifications. Based on the description above, the objective of this research was to determine the caffeine content in the bottle drink branded X, Y, and Z by HPLC method.

2. Materials and Methods

2.1. Material

All solvents and reagent were analytical grade, chloroform, Dragendorff reagents, HCl, cobalt chloride, double distilled water, cobalt nitrate, methanol for liquid chromatography, ethanol.

2.2. Apparatus

The apparatus was used in this was LC-20AT HPLC instrument with SPD-20A Shimadzu detector. HPLC conditions referred to previous study [8] with slight modification in mobile phase composition ie : reversed phase, C18 ODS columns 250 x 4.6 mm, detector set at 278 nm wavelength, flow rate 1 mL/min at room temperature, 20.0 μL of sample volume, and methanol motion phase: double distilled water (95:5).

2.3. Experimental

2.3.1 Sample collection

Samples were obtained from several stores which selling coffee packaging with initials X, Y, and Z which are the "3 top brand index phase 2".

2.3.2. Qualitative analysis

2.3.2.1. Dragendorff test

A total of 1.0 mL of the sample was dissolved in 10.0 mL of chloroform and 4 drops of NH_4OH, then filtered and the filtrate was inserted into a closed reaction tube. The chloroform extract in the test tube was shaken with 6.0 mL H_2SO_4 2M and the acid layer was separated into another test tube. This acid layer is dripped onto the drop plate and added Dragendorff reagent which will cause the orange red sediment [9].

2.3.2.2. Parry method

Parry method analysis was performed by means of a number of substances dissolved in alcohol, then added Parry reagents and dilute ammonia. The dark blue/green solution states that show caffeine exists (MOH, 1995). Parry reagents were made by reacting Cobalt Nitrate [$Co(NO_3)_2$] with methanol (CH_3OH)[10]

2.3.3. Quantitative analysis

The system suitability test was performed by injecting a standard solution of caffeine 60 μg/mL as 6 times in a predetermined HPLC system. The RSD value is specified for the values of t_R, AUC, and tailing factors value.

The linearity test was performed by injecting the caffeine standard soluble in ethanol with concentrations of 40, 60, 80, 100, 120, 140, 160, 180, and 200 μg/mL, respectively, into the HPLC system. The standard curve was created between the concentration of caffeine standard and AUC. [8].

In order to determine the caffeine content in coffee beverages, 5.0 mL coffee sample was inserted in a separating funnel, then extracted 4 times, each with 25.0 ml of chloroform addition. The bottom layer is taken, then the extract (chloroform layer) is evaporated over the water bath until the chloroform evaporates entirely. The solvent-free caffeine extract is dissolved with ethanol up to 5.0 mL [10].

The standard solution and sample were injected into HPLC system, with HPLC conditions as follows: reversed phase, C18 / ODS columns 250 x 4.6 mm, detector set at 278 nm wavelength, flow rate 1 mL/min at room temperature, sample volume 20, 0 μL, and mobile phase methanol: double distilled water (95:5). The caffeine content is calculated with the linear regression equation $y = bx + a$ from the standard calibration curve [11].

IOP Conf. Series: Materials Science and Engineering **259** (2017) 012011 doi:10.1088/1757-899X/259/1/012011

3. Results and Discussion

3.1. Qualitative analysis

3.1.1. Dragendorff test

The result of the qualitative test with Dragendorff test was presented in Table 1.

Tabel 1. The result of Dragendorff test on coffee beverages

Sample	Before added reagent	After added reagent	Result
Caffeine Standard	Clear	Brown precipitation	+
Sample "X"	Clear	Brown precipitation	+
Sample "Y"	Clear	Brown precipitation	+
Sample "Z"	Clear	Brown precipitation	+

Note: + = contained alkaloid

Table 1 showed that three samples reacted with Dragendorff reagent and formed a red brick/brown precipitation. The caffeine standard also showed a red brick precipitate with Dragendorff. This result showed that in the three samples contained alkaloid compounds [12]. Caffeine was a purin alkaloid compound in a plant.

3.1.2. Parry Test

The positive result of caffeine compounds in the sample is shown by the formation of a dark blue/green solution. In this test, the Cobalt (Co) ions in the reagent will form a green complex. The cobalt ion has positively charged and allow to bind to the nitrogen group present in the caffeine compound [11]. The result of the qualitative test with Parry method can be seen in Table 2.

Tabel 2. The result of Parry method test of coffee beverages

Sample	Before added reagent	After added reagent	Result
Caffeine Standard	Clear	Green	+
Sample X	Clear	Green	+
Sample Y	Clear	Green	+
Sample Z	Clear	Green	+

Table 2 showed that 3 samples which tested using Parry reagents produced a green color. The same result was shown by caffeine standard. This result indicates the presence of caffeine in the sample. Maramis was also found the same result in the study of [6] samples which contain caffeine compounds.

3.1.3. Retention Time (t_R)

Qualitative analysis of caffeine content with HPLC was done by comparing the retention time of caffeine standard (pure compound) with the sample. From the experiment, the average standard caffeine tR was 2.518 minutes and the mean t_R of X, Y, and Z samples were 2.508 minutes, 2.491 minutes, and 2.495 minutes, respectively. From the data, it is clear that t_R of standard caffeine and the three samples are relatively the same, so it can be concluded that three samples had caffeine content. The solvent, standard caffeine and samples chromatogram profiles X, Y, and Z are presented in Figure 1.

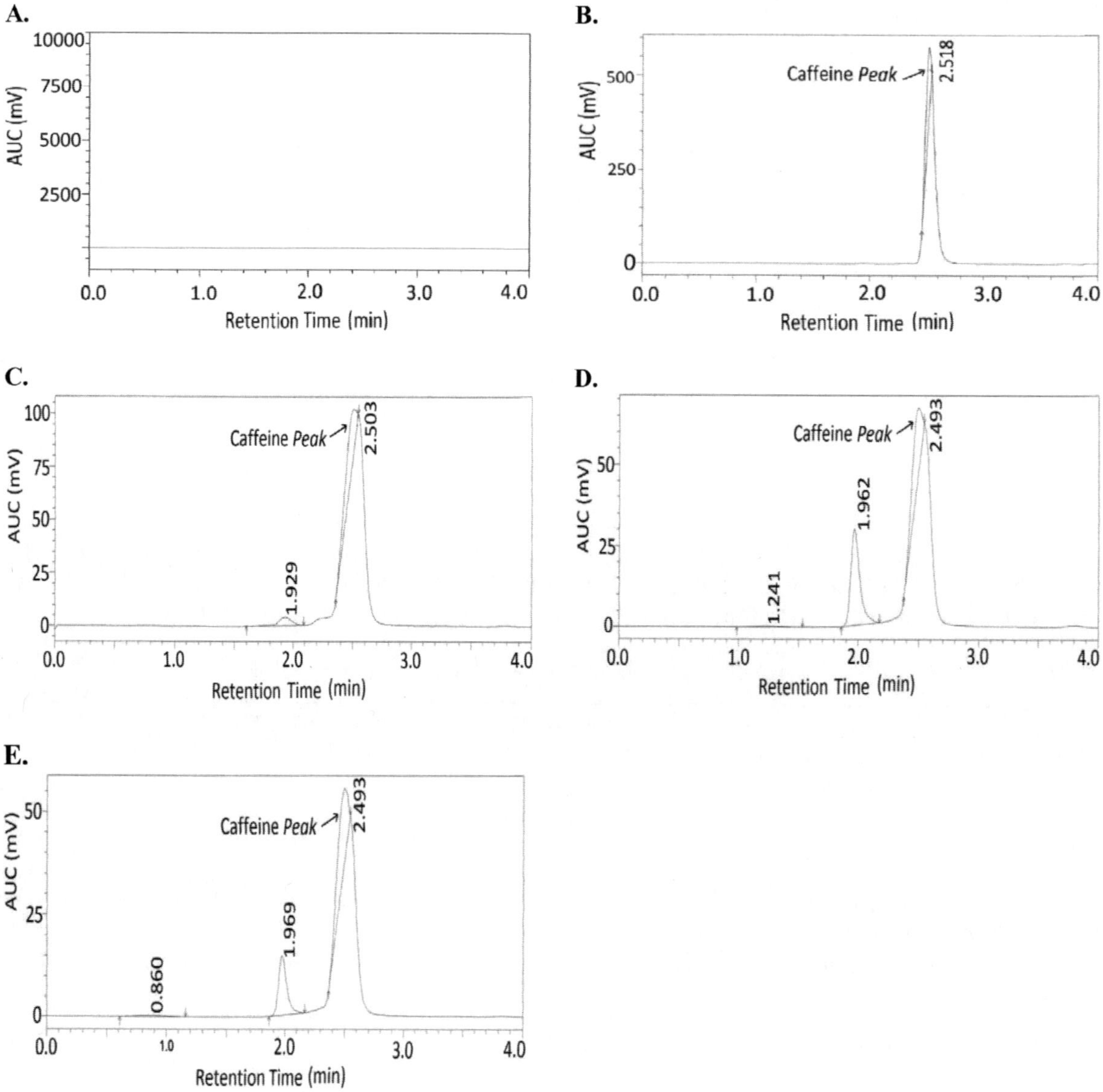

Figure 1. Chromatograms of (A) solvent, (B) caffeine standard solution, (C) sample X, (D) sample Y, and (E) sample Z

3.2. Quantitative Test

The result of system suitability test obtained RSD of tR equal to 0.199%, RSD of AUC equal to 1.476% and tailing factor equal to 0.67. These data showed that chromatographic system meets the requirements of RSD t_R and AUC < 2% and 0.5 < TF < 2[13]. The results showed a linear relationship between the concentration of standard caffeine solution with the concentration range of 40-200 μg/ml. the result was obtained linear regression equation $y = 15044x + 231298.87$ with the R value = 0.992 ($p < 0.05$) where X is the standard concentration of caffeine (μg/mL) and Y as AUC (mV).

Determination of caffeine content in coffee beverage samples
The caffeine content of the three coffee beverage samples as listed in Table 3 below.

Table 3. Caffeine levels in samples X, Y, dan Z

No	Caffeine Const. (mg/botle		
	X	Y	Z
1	133.159	106.953	148.797
2	130.683	109.057	146.292
3	132.969	109.583	145.541
4	132.969	112.038	148.547
5	131.445	110.810	150.049
6	131.064	109.758	146.793
mean	**132.048***	**109.699***	**147.669***
SD	**1.107**	**1.713**	**1.726**
%CV	**0.838%**	**1.562%**	**1.169%**

Note: * $p<0.05$

The caffeine content of the three coffee beverage samples are statistically different ($p<0.05$). According to Indonesian Standard Agency , the allowed caffeine levels are 150 mg/day (SNI 01-3542-2004) and 50 mg/serving (SNI 01-7152-2006) [4]. While based on the FDA (Food Drug Administration) [5] the dose of caffeine is allowed for adult <400 mg/day. However, the present study showed that caffeine content does not comply with the SNI 01-7152-2006, where in a serving should only contain a maximum of 50 mg of caffeine while in samples X, Y, and Z levels obtained are 132.048 mg/bottle; 109.699 mg/bottle, and 147.669 mg/bottle (Table 3). This level exceeds the limit set by Indonesian Standard Agency for one serving of coffee. However, based on SNI 01-3542-2004 the limit of caffeine allowed to be consumed is 150 mg/day, while the FDA regulation allowed the level of caffein <400 mg/day.

4. Conclusions

The caffeine content of the X, Y, and Z coffee beverage samples respectively were 138.048 mg/ bottle; 109.699 mg / bottle, and 147.669 mg / bottle.

The levels of caffeine contained in the X, Y, and Z coffee beverage are not comply with Indonesian Standard Agency regulation.

References

[1] Tjay H and Rahardja K *Essential Drug, It's Benefits, uses and Side effect* (Jakarta: Elexmedia Komputindo).
[2] Danasrayaningsih V 2007 Penetapan kadar kafein dalam minuman berenergi merk 'X' dengan metode spektrofotometri derivatif aplikasi peak to peak, Universitas Sanata Dharma.
[3] Freedman N D 2012 Association of coffee drinking with total and all cause mortality *Nejm*, 1–14.
[4] Arwangga A F, Asih I A and Sudiarta I W 2016 Analisis kandungan kafein pada kopi di desa sesaot narmada menggunakan spektrofotometri Uv-Vis *J. Kim. Univ. Udayana* **10** (1) 110–114.
[5] FDA FDA to Investigate Added Caffeine 2013 [Online]. Available: https://www.fda.gov/ForConsumers/ConsumerUpdates/ucm350570.htm.
[6] Fatoni A 2015 Analisis secara kualitatif dan kuantitatif kadar kafein dalam kopi bubuk lokal yang beredar di kota Palembang menggunakan spektrofotometer UV-Vis. Palembang.

[7] Aptika N, Tunas I and Sutema I 2015 Analisis kadar kafein pada kpi hitam di Lebah Bukian Gianyar menggunakan spektofotometri UV-Vis *J. Chem. Lab.* **2** (1) 30-37.
[8] Wanyika H N, Gatebe E G, Gitu L M, Ngumba E K, and Maritim C W 2010 Determination of caffeine content of tea and instant coffee brands found in the Kenyan market **4** 353-358.
[9] Marliana S D and Suryanti V 2005 Skrining fitokimia dan analisis kromatografi lapis tipis komponen kimia buah labu siam (*Sechium edule* Jacq. Swartz.) dalam ekstrak etanol *Biofarmasi* **3** (1) 26–31.
[10] Maramis R K, Citraningtyas G and Wehantouw F 2013 Analisis kafein dalam kopi bubuk di kota manado menggunakan spektrofotometri Uv-Vis *J. Ilm. Farm.***2** (4) 122-128.
[11] Artanti A N and Nikmah W R Setiawan H and Prihapsara F 2016 Perbedaan kadar kafein daun teh (*Camellia sinensis* (L.) Kuntze) berdasarkan status ketinggian tempat tanam dengan metode HPLC *J. Pharm. Sci. Clin. Res.* **1** (1) 37–44.
[12] Kusrahman A 2012 Isolasi, karakterisasi senyawa aktif dan uji farmaka ekstrak biji kebiul pada mencit Universitas Bengkulu.
[13] Snyder L R, Kirkland J J and Dolan J W *Introduction to Modern Liquid Chromatography, Third Edition*, Third. **22** (1) (USA: John Willey and Son's).

The influence of black cumin seed oil therapy with dosage of 1.5 mL/day and 3 mL/day to interleukin-21 (IL-21) expression of the patients with metabolic syndrome risk

D R Fajar[1], Akrom[1,2], E Darmawan[1]

[1]Faculty of Pharmacy , Universitas Ahmad Dahlan, Yogyakarta, Indonesia

[2]Center of Drug Information and Study, Universitas Ahmad Dahlan, Yogyakarta, Indonesia

E-mail: desi.rf1991@gmail.com

Abstrak. Metabolic syndrome (MS) is a metabolic disorder caused by obesity and insulin resistance. In the state of obesity and insulin resistance occurs increased fat metabolism that causes production reactive oxygen species (ROS) and oxidative stress that makes dysregulation of adipose tissue that decreases antioxidant enzymes and immune system disorders. In diabetes mellitus there is pancreatic β cell damage caused by pro-inflammatory cytokines i.e interleukin-21 (IL-21). Black cumin seed oil (BCSO) contains antioxidants and immunomodulators but has not seen how it affects to IL-21. This study used cross over design to determine the effect of BCSO with dose of 1.5 mL/day and 3 mL/day dose of IL-21 in MS patients. The total subjects 66 of MS patients were divided into 2 groups. Group I was administered 1.5 mL/day and 3 mL/day of BCSO in dosage for 20 days and continued with washing out (BCSO discontinued) for 7 days. On the day 28, group I administrated 3 mL/day of BCSO and group II were given 1.5 mL/day of BCSO during 20 days. Statistical analysis showed that mean value expression of IL-21 were given by BCSO at dose 1.5 mL/day was 5.06±5.48 and BCSO at dose 3 mL/day 4.66±3.63 ($p<0.05$). The result showed that IL-21 expression in patient at risk of MS who received BCSO dose 3 mL/day were lower than those received adose of 1.5 mL/day for 20 days.

Keywords: black cumin seed oil, interleukin-21, metabolic syndrome risk

1. Introduction

Metabolic syndrome (MS) is a set of abnormalities involving various physiological, biochemical, clinical and metabolic factors that can lead to the development of cardiovascular disease and diabetes mellitus and all-cause mortality risk [1]. Some of the risks causing MS include lifestyle (diet, smoking, physical activity), genetics and socioeconomics [2]. MS in Indonesia is one of the main public health problems such as diabetes mellitus, central obesity, dyslipidemia, and hypertension, which is a component of MS and has an effect on the prevalence of metabolic syndrome. MS case rate in Indonesia reached 23% i.e 26.6% in women and 18.3% in men of productive age, this data is based on the result of basic health research [3].

The pathophysiology of metabolic syndrome is not known for certain, but some literature say that the primary reason is a person's obesity state. Excess fat in the body causes increased fat metabolism along with reactive oxygen species (ROS) production as well as increased oxidative stress characterized by a balance disorder of oxidation reduction reactions and decreased antioxidant enzymes in both adipose and tissue cells, this situation occurs in hypertension and atherosclerosis [4].

In addition, fat cell enlargement will secrete metabolic products such as pro-inflammatory cytokines, procoagulants, inflammatory peptides and angiotensinogen [5]. Insulin resistance affects type 2 of diabetes and hyperglycemia. A person with diabetes, oxidative stress will inhibit the taking of glucose in muscle cells and fat cells and decrease insulin secretion by pancreatic β cells (type 2 diabetes). This is due to autoimmune disorders mediated by lymphocytes T and activated dendric cells involving inflammatory pathways Thelper1 (Th1), Thelper2 (Th2) and Thelper17 (Th17) and inhibition of T regulators. It is found that the Th17 pathway is involved in the inflammatory process and damage to pancreatic beta cells regulated by interleukin-21, interleukin-23 and interleukin-27. Interleukin-21 is a type 1 of cytokine [6, 7].

BSCO contains antioxidant and immunomodulatory substances which can be used as antidiabetes mellitus [8]. The main content of BSCO is essential oils such as timokuinon that have an effect on the immune system through inhibition of translocation of NF-kB into the nucleus so as to decrease pro-inflammatory cytokines, chemokines, adhesion factors and coagulation factors[8]. In addition, it is known that BSCO can decrease the oxidative stress condition that triggers the synthesis of pro-inflammatory cytokines through adhesion molecules. Interleukin-21 is a pro-inflammatory cytokine that plays a role in pancreatic B cell damage but it is not yet known how the effect of BSCO with a dose of 1.5 mL/day and 3 mL/day for 20 days in patients with metabolic syndrome risk.

2. Materials and Methods

This study used cross over design with a total of 66 patients with metabolic syndrome who state willing to follow the study by agreeing inform consent. The inclusion criteria were obesity, blood pressure ≥140/90 mmHg, triglyseride level ≥150 mg/dL or using the drug of methabolic syndrom and 18-70 years old. Exclusion criteria werepregnant and lactating women, tuberculosis and be trought immun disease. But during the course of the study there were 4 patients who were excluded due to side effects such as nausea, vomiting, gastritis, increased blood pressure and decreased kidney function. This research was conducted at Jetis I Public Health Centers, Bantul Regency, Yogyakarta in August 2016 until March 2017. This clinical trial protocol has been reviewed and approved by health research ethics committee of Faculty of Medicine and Health Sciences, Universitas Muhammadiyah Yogyakarta with approval number 279/EP-FKIK-UMY/VIII/2016.

Blood steering patients were taken from vena cubity by analysts in laboratory of the Jetis I Public Health Centers, Bantul Regency, Yogyakarta. Procedures undertaken in the implementation have been reviewed by research ethics committee of Universitas Muhammadiyah Yogyakarta. Blood that has been accommodated in a vacutainer containing anticoagulants is taken to the pharmacology laboratory of Universitas Gadjah Mada (UGM) and ready to be examined with flow cytometer. Here is the scheme of research implementation:

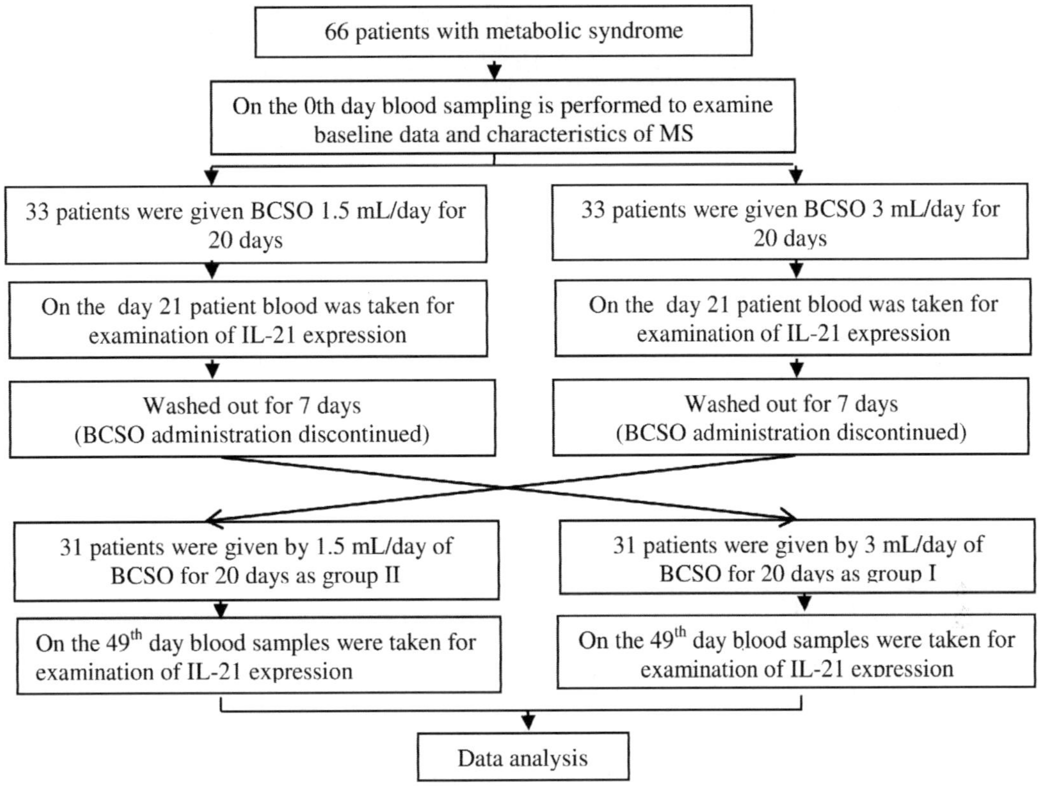

Figure 1. Schematic of research design

3. Results and Discussion

3.1. The characteristics data of patients at risk of metabolic syndrome at Jetis I Public Health Centers, Bantul Regency in 2016

From the previous studies, demographic data and characteristic data of metabolic syndrome as primary outcome and percentage of IL-21 expression as outcome secunder. From the data, statistical test using Chy-Square test and paired T-test sample with 95% confidence level.

This research is conducted at Jetis 1 Public Health Centers, Bantul Regency, Yogyakarta in August 2016 - March 2017. We involved doctors in getting patients who will be joined in the research and obtained a total of 66 patients at risk of metabolic syndrome who have met the inclusion and exclusion criteria. However, during the study there were 4 patients who drop out because they complain the side effects such as nausea, vomiting, gastritis, increased blood pressure and decreased kidney function. The total subjects who followed the study to the end were 62 patients at risk of metabolic syndrome.

Table 1 shows the characteristics of patients at risk of metabolic syndrome. Based on gender, it is known that the metabolic syndrome was mostly suffered by women (78.80%) compared to men (21.20%). Metabolic syndrome was also more prevalent in the age group 51-60 years (42.40%), married status (98.50%), high school education (78.80%) and employment status (57.60). This result is in line with research conducted before [9] which reported that there was a significant relationship among gender, age, and work with the incidence of metabolic syndrome. Metabolic syndrome is more risky in women than in men. In women there is an excessive accumulation of fat in the adipose tissue that causes hypertriglyceride. Metabolic syndrome in women is also caused by diet, genetics and lack of physical activity than men [9, 10]. Women are more at risk of developing diabetes mellitus due to body mass index (BMI) that is more easily increased and the distribution of body fat easily accumulated due to hormonal processes in the body [11].

Table 1. The characteristics data of patients at risk of metabolic syndrome at Jetis I Public Health Centers in 2016

Characteristics of patients	n=66	Percentage (%)
Gender		
Men	14	21.20
Women	52	78.80
Age		
<41 years old	1	1.50
41-50 years old	20	30.30
51-60 years old	28	42.40
>60 years old	17	25.80
Marital status		
Married	65	98.50
Single	1	1.50
Education		
No school	6	9.10
Elementary-High School	52	78.80
College-Master	8	12.10
Working		
Work	38	57.60
Not working	28	42.40

The incidence of metabolic syndrome is more experienced by age group > 40 years than age < 40 years [9]. Metabolic syndrome occurs as one gets older. This is due to a decrease in organ function and metabolic disorders that exist in the body. Diabetes mellitus occurs due to decreased insulin secretion because of the pancreatic β cells damage, insulin resistance, impaired metabolism of carbohydrates, proteins and fats and reduced elasticity of blood vessels may be at risk of hypertension and atherosclerosis leading to the incidence of metabolic syndrome [9, 12, 13]. Metabolic syndrome has a significant relationship with one's marital status [14, 15]. A married person will be faced with a social environment that allows problems and causes unhappiness. Each problem will create psychosocial stress potential that may be at risk of metabolic syndrome [15]. Education and work are two characteristics of metabolic syndrome which has interconnected each other. Education can determine the type of work of a person with different income to meet their daily needs. Lifestyle change can increase the prevalence of diseases such as diabetes, obesity and heart disease which is a factor risk of metabolic syndrome [16]. Job status relates to daily physical activity. In this study, patients with metabolic syndrome are more likely than workers, such as civil servants, entrepreneurs, farmers, fishermen and laborers with mild to severe activity. Physical activity can help lower cholesterol, triglycerides, increase HDL and insulin sensitivity and lower blood pressure which is a risk factor for the incidence of metabolic syndrome and cardiovascular disease complications [17].

Table 2 shows the characteristics data of patients with metabolic syndrome. Metabolic syndrome is more prevalent in patients at risk of metabolic syndrome with prolonged <5 years (59.10%) compared with 5 years of age (40.90%). Patients at risk for metabolic syndrome who participate in this study mostly suffered from diabetes mellitus with complications. Diabetes mellitus is a metabolic disease with characteristics of hyperglycemia due to abnormalities of insulin secretion and become a risk factor for organ damage and dysfunction especially in the eyes, kidneys, nerves, heart and blood vessels.

Table 2. The characteristics data of patients at risk of metabolic syndrome at Jetis I Bantul Public Health Centers in 2016

Characteristics of patients	n=66	Percentage(%)
Characteristics of metabolic syndrome		
DM	10	15.20
Hypertension	1	1.50
Hypertriglyceride	1	1.50
DM+ Hypertension	16	24.20
DM+ Hypercholesterolemia	2	3.00
DM+ Obesity	1	1.50
DM+ Hypertriglyceride	6	9.10
DM+ Dyslipidemia	3	4.50
DM+ Hypertension + Hypertriglyceride	12	18.20
DM+ Hypertension + Hypercholesterolemia	5	7.60
DM+ Hypertension + Obesity	2	3.00
DM+ Hypertension + Dyslipidemia	5	7.60
DM+ Hypertension + Hypertriglyceride + Obesity	2	3.00
Routine Medication		
DM Drug	25	37.90
Hypertension Drug	1	1.50
Dyslipidemia Drug	1	1.50
DM Drug+ Hypertension Drug	33	50.00
DM Drug +Dyslipidemia Drug	3	4.50
DM Drug + Hypertension Drug + Dyslipidemia Drug	3	4.50

3.2. Clinical examination result of patients at risk of metabolic syndrome before and after using BCSO Dose 1.5 mL/day and 3 mL/day

Patients at risk of metabolic syndrome before treatment are examined clinical conditions such as systolic and diastolic blood pressure, pulse rate, body mass index (BMI), using a tool that has been calibrated by the calibration laboratory of Universitas Ahmad Dahlan Yogyakarta. Examination of blood glucose level (BG), triglyceride level (TG) and SGOT and SGPT activity using 5010 spectrophotometer at the laboratory of Jetis I Public Health Centers, Bantul Regency, Yogyakarta. The data were analyzed using repeated measure ANOVA test and continued with Bonferroni test as shown in Table 3.

Table 3. The results of measurement of clinical condition of patients at risk of metabolic syndrome before and after using BCSO dose 1.5 mL/day and 3 mL/day at Jetis I Public Health Centers, Bantul Regency, Yogyakarta in 2016

Clinical Conditions of Patients	Average ± SD, n=62			p Values
	Pretreatment	Dose 1.5 mL/day	Dose 3 mL/day	
SBP(mmHg)	143.06±17.98[a]	137.73±17.66	136.92±17.66	0.00
DBP(mmHg)	80.78±9.25	79.27±13.96	79.95±10.98	0.58
Pulse (x/min)	88.66±13.82[a]	90.69±12.91	92.44±12.15	0.04
BMI (kg/m^2)	24.16±4.13[a]	24.79±4.27	24.35±5.10	0.00
BG (mg/dL)	246.69±109.30[ab]	184.18±89.46	197.77±87.14	0.00
TG (mg/dL)	193.84±114.61[a]	173.42±108.51	158.58±97.87	0.00
SGOT (U/L)	23.00±10.41	21.82±6.47	22.60±9.95	0.63
SGPT (U/L)	22.29±10.81	21.16±7.86	21.03±12.08	0.82

Description: a= there is a significant difference between the measurement before treatment and the dose measurement 3 mL/day; B= there is a significant difference between the measurement before treatment and the dose measurement of 1.5 mL/day.

The results of statistical analysis show that there are significant differences between measurement of systolic blood pressure (SBP), pulse, BMI, BG and TG before and after BCSO administration in patients at risk of metabolic syndrome ($p<0.05$). The results of statistical analysis show that there is no significant difference between diastolic blood pressure (DBP) measurement, SGOT and SGPT activities before and after BCSO administration in patients at risk of metabolic syndrome ($p>0.05$). These results indicate that giving BCSO dose of 1.5 mL/day and 3 mL/day for 20 days can decrease the mean blood pressure value in patients at risk of metabolic syndrome with targeted therapy <140/90 mmHg. This is in line with research conducted which states that the provision of black cumin seed powder can reduce blood pressure during 20 days of administration in patients with metabolic syndrome [18]. Another study mentioned that giving black seeds of cumin seeds for 8 weeks can reduce the blood pressure of cytolic and diastolic without side effects [19]. This study shows that black seeds of cumin seeds for 20 days are able to increase pulse rate and BMI in patients at risk of metabolic syndrome but still within normal limits. According to the American Heart Association (AHA) the normal range of pulse rate is 60-100 x/min and according to the World Health Organization (WHO) the normal value of IMT in patients at risk of metabolic syndrome is <30 kg/ m2. The results of this study stated that the administration of BCSO dose 1.5 mL/day and 3 mL/day for 20 days can reduce blood glucose levels in time and triglycerides in patients at risk of metabolic syndrome. BCSO contains active compounds such as thymoquinone, nigelon and thymol which play a role in lowering blood glucose. Black cumin in adipose cells can stimulate the secretion of insulin produced by pancreatic cells and is shown to have no effect on tissue sensitivity to glucose [20]. Criteria for metabolic syndrome according to International Education DiabetesFederation (IDF) in hypertension, triglycerides >150 mg/dL, HDL for women < 50 mg/dL and men < 40 mg/dL or in treatment of tubercidemia, diagnosed type 2 DM with GDP > 100 mg/dL [21].

Statistical analysis shows that SGOT and SGPT activities in patients at risk of metabolic syndrome decrease after using BCSO dose 1.5 mL/day and 3 mL/day for 20 days but not statistically significant ($p>0,05$). This is because the patients who participate in this study had a mean value of SGOT and SGPT activity in normal limits. The incidence of metabolic syndrome is associated with fatty liver disease or NAFLD. In patients with insulin resistance will experience steatosis due to increased free fatty acids to the liver increased. Oxidative stress and cytokines can cause disruption to the function of cells and liver organ [22]. SGOT and SGPT can be an indicator of liver damage with SGOT values <35 U/L and SGPT <41 U/L.

3.3. Measurements of IL-21 expression in patients at risk of metabolic syndrome given BCSO dosage 1.5 mL/day and 3 mL/day

Measurements of IL-21 expression using a flow cytometer tool in the pharmacology laboratory of the Faculty of Medicine, Universitas Gajah Mada (UGM), Yogyakarta. Data on IL-21 expression measurements in patients at risk of metabolic syndrome given BCSO dose 1.5 mL/day and 3 mL/day for 20 days statistically analyzed using paired T-test as shown in Table 4.

Table 4. The results of IL-21 expression measurements in patients at risk of metabolic syndrome given BCSO dose 1.5 mL/day and 3 mL/day for 20 days

Clinical condition patient	Mean value ± SD		p Value
	Dose 1.5 mL/day (n=62)	Dose 3 mL/day (n=62)	
IL-21	5.06±5.48	4.66±3.63	0.69

The results of statistical analysis shows that the mean value of IL-21 expression in patients at risk of metabolic syndrome who received BCSO dose 3 mL/day was lower than those who received BCSO dose 1.5 mL/day but not statistically significant ($p>0.05$). Several factors that may influence the research results are the possibility of influence from previous treatment before cross over design on BCSO dose 1.5 mL/day and 3 mL/day for 20 days. Physiological conditions of the body and lifestyle such as lack of exercise, smoking and taking drugs for long periods in patients at risk of metabolic syndrome can also affect the results of the study.

Metabolic syndrome occurs because of insulin resistance. This is due to the disruption of pancreatic β cell function and insulin hypersecretion resulting in insulin resistance, in addition to severe damage to pancreatic β cells that leads to decreased insulin production and hyperglycemia. The deficiency of IL-21 expression in diabetic nonobese mice (NOD) may lead to resistance to insulitis, insulin autoantibody production and the onset of type I diabetes [7]. Excess of IL-21 expression in pancreatic β cells makes of inducing chemokines, inflammatory cytokines such as IL-17A, IL-17F, IFN-gamma, monocyte chemoattractant protein (MCP)-1 and MCP-2. IL-21 is a type I cytokine produced by T-cells and NK cells and is detected in $CD4^+$ and T-helper undergoing differentiation [23].

Giving *Nigela sativa* L. can cause demodulator in pro-inflammatory cytokines produced by T-helper-1 and cytokines. The active substance contained in the BSCO can evoke the immune system with the ability to raise T-helper cell, Tsupressor cells, and NK cells are a product of lymphocytes [24, 25]. T-cells play an important role in inducing inflammation induces makrofaf and secrete inflammatory and pro-inflammatory cytokines in adipose tissue leads to insulin resistance and lead to diabetes mellitus. Inflammatory cytokines would improve insulin resistance through the role of inflammatory signals such as janus kinase (JK) and activation of transcription factors nuclearfactor kappaB (NF-kB). Signaling pathways such as JK and NF-kB may reduce the action of insulin in adipocyte tissue [26, 27].

Black cumin seed can be utilized as an antioxidant and immunomodulator. Hexane extracts of black cumin seeds (*Nigella sativa* L.) can improve macrophage phagocytic activity in DMBA-induced female rats in vitro [8]. Part of the most widely used black cumin seed plant is its oil. Black cumin seed oil contains essential oil content of non polar thymoquinone that is 27.8% [28]. The working mechanism of thymoquinone is by inhibiting non-enzymatic lipid peroxidation in liposomes and has the potential to neutralize free radicals and inhibit the pro-inflammatory cytokines such as interleukin-1β, interleukin-8 and chemokines such as machrophage chemotactic protein (MCP)-1 with the mechanism of action via inhibition of TNF- α which induces activation of transcription factors nuclearfactor kappaB (NF-kB) signaling pathway is involved in IL-21 and inhibiting the translocation of NF-kB into the nucleus [29]. BCSO delivery in the metabolic syndrome will inhibit IL-21 activation by decreasing the deference of T-helper which also produces IL-21 (IL-21R) receptors as well as inhibition of the STAT 3 signaling pathway.

4. Conclusion

IL-21 expressions in patients at risk of metabolic syndrome who received BCSO dose 3 mL/day were lower than those received a dose of 1.5 mL/day for 20 days.

References

[1] Kaur J 2014 A comprehensive review on metabolic syndrome, Cardiology Research and Practice, ID 943162.

[2] Sandra R 2015 Sindrom metabolik *Jurnal Majority* **4** (4) 88, Fakultas kedokteran Univeristas Lampung.

[3] Ministry of Health 2014 Riset Kesehatan Dasar (RISKESDAS 2013), Jakarta.

[4] Stocker R and Keaney JF 2004 Role of oxidative modification in atheroclerosis *J. Physiol.* **84** (5) 1381-1392.

[5] Semiardji 2004 The significant of visceral fat in metabolic syndrome, Jakarta: *Diabetes meeting* 9-10 Oktober.

[6] Dandona P, Aljada A and Chaudhuri A 2005 Metabolic syndrome: A comprehensive perspective based on interaction between obesity and diabetes *Circulation* **111**(11):1448-54.

[7] Andrew P, Tom, V B, Andrea L, Suto Akira, Michaud M, Zhang D, Grusby M and Herrath M V 2009 Interleukin-21 is required for the depelopment of type 1 diabetes in NOD Mice *Diabetes* **58**(5) 1144-1155.

[8] Akrom, 2013, Mekanisme kemoprefentif ekstrak heksan BJH pada tikus SD diinduksi DMBA: kajian antioksidan dan immunomodulator *Disertasi* Program Doktor Ilmu Kedokteran dan Kesehatan, FK UGM, Yogyakarta.
[9] Suhaema and Masthalina H 2015 Pola Konsumsi dengan Terjadinya Sindrom metabolik di Indonesia *Jurnal Kesehatan Masyarakat Nasional* **9** 4.
[10] Chen H 2006 Enhancing energy and glucose metabolism by distrupting trigliseride syntesis: lesson from Mice Lacking DGAT-1 *Nutrition & Metabolism* **3** 10.
[11] Irawan D 2010 Prevalensi dan faktor resiko kejadian diabetes melitus tipe 2 di daerah Urban Indonesia *Tesis* Universitas Indonesia; Jakarta.
[12] Jelantik I M and Haryati E 2014 Hubungan faktor resiko umur, jenis kelamin, kegemukan, hipertensi dengan kejadian diabetes mellitus tipe 2 di wilayah kerja Puskesmas Mataram *Media Bina Ilmiah* **8** 1 39-44.
[13] Zahtamal, Chandra F, Suryanto and Restuastuti T 2007 Faktor pasien diabetes mellitus, *Berita Kedokteran Masyarakat* **23** (3) 142-147.
[14] Mohammad A G, Abdulbari B, Asia N, Ismail, Mahmoud Z, Waleed K, Abdullatef, Abdullah OAA and Al Hamaq 2010 Gender and age related differences in patients with the metabolic syndrome in a highly endogamus population *J. Basic Med. Sci.* **10** (3) 210-217.
[15] Troxel K A, Matthews L C, Gallo L H and Kuller 2005 Marital quality and occurrence of the metabolic syndrome in women *Archives of Int. Med.* **165**(9) 1022-1027.
[16] Praditwi R 2008 Faktor risiko yang mempengaruhi terjadinya penyakit sindrom metabolik, *Skripsi,* Fakultas kesehatan masyarakat, Universitas Airlangga; Surabaya.
[17] Berttrais S, Beyeme JP, Czernichow S, Galan P, Hercberg S and Oppert JM 2005 Sedentary behaviors, physical avtivity and metabolic syndrome in middle-aged French subject *Obesity Res.* **13** 936-944.
[18] Najmi A, Haque S F, Naseeruddin M and Khan R A 2008 Effect of *Nigella sativa* oil on various clinical and biochemical parameters of metabolic syndrome *Int. J. Diabetes & Metab.* **16** 85-87.
[19] Husaeni H F, Amini M, Mohtashami R, Ghamarchehre M E, Sadeqhi Z, Kianbakht S, Husein A F 2013 Blood pressure lowering effect of *Nigella sativa* L. seed oil in healthy volunteers: A randomized, double-blind, placebo-controlled clinical Trial *J. Phytother. Res.* **21** 12.
[20] Andaloussi AB, Martineau LC, Spoor D,Vuong T, Leduc C, Joly E 2008 Antidiabetic activity of *Nigella sativa* seed extract in cultured pancreatic-β cells, skeletal Muscle cells and adipocytes *Pharmaceut. Biol. J.* **49** 96-104.
[21] Mittal S 2008 *The Metabolic Syndrome in Clinical Practice* (London: Springer-Verlag).
[22] Medina J, Fernandez S, Garcia B, and Moneno O 2004 Approach the patogenesis and treatment of nonalcoholic steatohepatitis *Diabetes care* **27** 2057-2066.
[23] Monteleone G, Pallone F and Macdonald TT 2009 Interleukin-21 as a new therapeutic target for immune-mediated diseases, *Trends in Pharmacological Sciences* **30** 8.
[24] Islam and Muhammad T 2016 Biological avtivities and therapeutic promises of *Nigella sativa* L *Int. J. of Pharma Sci. and Scientific Res.* **2** 6 242-243
[25] Gilani AH, Jabeen Q and Khan MAU, 2004 Review of medical uses and pharmacological activities of *Nigella sativa* L seeds *Pak. J. Biol. Sci.* **4** 441-451.
[26] Arkan M C, Hevener A L, Greten F R, Maeda S, Li Z W and Long J M 2005 IKK beta links inflammation to obesity-induced insulin resistance *Nat. Med.* **11**:191.
[27] Hirosumi J, Tuncman G, Chang L, Gorgun CZ, Uysal KT and Maeda K 2002 A central role for JNK in obesity and insulin resistance *Nature J.* **6**:333.
[28] Claudia C, Maria G, Hanganu D, Olah N, Maria F, Hammam C and Hammam M 2010 Chemical composition of the tunisian *Nigela sativa* Note I. Profile on essential oil *Farmacia* **58** 4.

IOP Conf. Series: Materials Science and Engineering **259** (2017) 012012 doi:10.1088/1757-899X/259/1/012012

[29] Ricky TN 2011 Pengaruh pemberian terapi tambahan adjuvan minyak jinten hitam (*Nigella sativa)* terhadap hitunglLimfosit mencit Balb/C model sepsis *Skripsi,* Fakultas Kedokteran Universitas Sebelas Maret: Surakarta.

IOP Conf. Series: Materials Science and Engineering **259** (2017) 012013 doi:10.1088/1757-899X/259/1/012013

Direct medical cost and utility analysis of diabetics outpatient at Karanganyar public hospital

Eristina[1], **T M Andayani**[2], **R A Oetari**[1]

[1]Faculty of Pharmacy, Setia Budi University, Surakarta, Indonesia

[2]Department of Pharmacology and Clinical Pharmacy, Faculty of Pharmacy, Universitas Gadjah Mada, Yogyakarta, Indonesia

E-mail: ngaheristina@gmail.com

Abstract. Diabetes Mellitus is a high cost disease, especially in long-term complication treatment. Long-term complication treatment cost was a problem for the patient, it can affect patients quality of life stated with utility value. The purpose of this study was to determine the medical cost, utility value and leverage factors of diabetics outpatient. This study was cross sectional design, data collected from retrospective medical record of the financial and pharmacy department to obtain direct medical cost, utility value taken from EQ-5D-5L questionnaire. Data analyzed by Mann-Whitney and Kruskal-Wallis test. Results of this study were IDR 433,728.00 for the direct medical cost and pharmacy as the biggest cost. EQ-5D-5L questionnaire showed the biggest proportion on each dimension were 61% no problem on mobility dimension, 89% no problems on self-care dimension, 54% slight problems on usual activities dimension, 41% moderate problems on pain/discomfort dimension and 48% moderate problems on anxiety/depresion dimension. Build upon Thailand value set, utility value was 0.833. Direct medical cost was IDR 433,728.00 with leverage factors were pattern therapy, blood glucose level and complication. Utility value was 0.833 with leverage factors were patients characteristic, therapy pattern, blood glucose level and complication.

Keywords: diabetes mellitus, direct medical costs, utility value, EQ-5D-5L

1. Introduction

Diabetes mellitus (DM) is a metabolic disorder which is marked with hyperglycemia and anomaly in carbohydrate, fat and protein [1]. DM is diseases that require a lot of therapy costs, particularly long–term costs. Long–term cost treatment is a burden to the patients of diabetes mellitus, which affects the quality of life of the patients. Patients quality of life can be seen from the utility value. Utility is the satisfaction of the patients after getting health intervention [2]. EQ-5D-5L questionnaire used on this study to measure patient's utility. EQ-5D has the advantage to measure the single comprehensive score of the unit of the value which is able to be compared among disease and used for economic evaluation [3].

2. Materials and Methods

This study was an analytic research with cross sectional design that is a research done in one time and at a time. Data taken in 2 ways, used EQ-5D-5L questionnaire and medical record of financial and pharmacy department. Data analyzed by calculating direct medical cost and utility value in conjuction with seeing relationship toward factor which may affect. Direct medical costs used in this study were:

administrative costs, medical fees, laboratory fees and drug costs and utility was patient satisfaction level after obtaining a health intervention that is measured by using EQ-5D-5L questionnaire. Questionnaire of EQ-5D-5L consists of 2 parts, the first describes a system consisting of 5 dimensions (mobility, self-care, activity, pain/discomfort, anxiety/depression). Second was the best and worst health status Visual Analogue Scale (VAS) 100 and 0 respectively. Respondents were asked to determine health status at which point and what line. An index utility with a range of 0 shows death and 1 represents perfect health. The value of index utility obtained by summing the result of the domain based on the value set of Thailand.

Accordance to hospital rules, 100 patients were used as sample by quota sampling technique. Inclusion criteria were outpatient who had undergone at Karanganyar Public Hospital for at least 3 months, outpatients who visited in January 2017 until February 2017 at Karanganyar Public Hospital, outpatients of type 2 DM both male or female >18 years, outpatient who willing to fill out the EQ-5D-5L questionnaire, complete medical record data include administrative costs, medical fees, laboratory fees and drug costs. Exclusion criteria were outpatients with mental disorders or language barriers that can disrupt the process of research. This study used hospital perspective, cost calculated was direct medical cost.

3. Results and Discussion

3.1. Direct medical cost

Result of direct medical cost components refer pharmacy as the highest component's cost (81.37%) caused by antidiabetic therapy and complication treatment which made pharmacy's cost higher than other components, shown in Table 1. The same result which medicine cost (53.27%) was the highest cost [4].

Table 1. Direct medical cost of DM outpatient at Karanganyar public hospital

Component's cost	n (IDR)	Average (IDR)	%
Pharmacy	35,294,037.00	352,940.00 ± 251,646.55	81.37
Laboratory	6,778,829.00	67,788.00 ± 60.527.19	15.63
Administration	1,200,000.00	12,000.00 ± 0	2.77
Services	100,000.00	1,000.00 ± 0	0.23
Total	43,372,866.00	433,728.00 ± 192,982.56	100

This study found average combination of insulin antidiabetic therapy pattern was IDR 1,069,185.71 which was higher than others pattern. The same result of insulin antidiabetik is bigger than the other antidiabetik, due to the price of insulin is higher that the other [5]. Insulin antidiabetic cost were 3 times greater than oral antidiabetics [6]. The result of SPSS Kruskal-Wallis Test ($p=0.000$) there was a significant relationship ($p<0.05$), therapy pattern has relation to direct medical cost, shown in Table 2.

Table 2. Therapy pattern and direct medical cost

Therapy Pattern	n	Average (IDR)	*P*
Single oral antidiabetic	29	183,665.69 ± 77,087.94	0.000
Combination oral antidiabetic	27	425,267.96 ± 103,454.91	
Single insulin antidiabetic	17	400,397.94 ± 132,543.02	
Combination insulin antidiabetic	7	1,069,185.71 ± 53,551.40	
Oral insulin antidiabetic	20	613,663.05 ± 147,855.44	

Table 3 shown average of uncontrolable blood glucose level cost was IDR 533,236.48 higher than controlable blood glucose level cost. The result of Mann-Whitney test ($p=0.000$) there was significantly difference of utility value ($p<0.05$), hence concluded glucose level influence to direct medical cost. Controlable blood glucose level use single antidiabetic therapy pattern while uncontrolable blood glucose level use combination antidiabetic therapy pattern [7, 8].

IOP Conf. Series: Materials Science and Engineering **259** (2017) 012013 doi:10.1088/1757-899X/259/1/012013

Table 3. Blood glucose level and direct medical cost

Blood Glucose Level	N	Average (IDR)	*p*
Controlable	46	316,915.13 ± 170,766.20	0.000
Uncontrolable	54	533,236.48 ± 274,449.14	

Average complication cost was IDR 553,083.59 which higher than non complication cost, shown in Table 4. Result of SPSS Mann Whitney Test (p=0.001) there was a significant relationship ($p<0.05$). The complication which is occured has an impact to the cost [6].

Table 4. Complications and direct medical cost

Complications	n	Average (IDR)	*p*
Complication	32	553,083.59 ± 270,353.55	0.001
Non Complications	68	377,561.63 ± 229,552.30	

3.2. Utility

This study used EQ-5D-5L questionnaire. Questionnaire consists of 2 parts. The first part describes the patient's condition consisting of 5 dimensions of mobility, self-care usual activities, pain/discomfort and anxiety/depression. Second part were Visual Analogue Scale to find out the health status of today, patient was required to cross the scale between 0 to 100.

Description

EQ-5D-5L questionnaire used to measure health status of DM outpatient at Karanganyar. The results obtained from diabetic outpatients in Karanganyar Public Hospital who do not have a serious problem, shown in Table 5. 61% no problem on mobility dimension, 89% no problems on self-care dimension, 54% slight problems on usual activities dimension, 41% moderate problems on pain/discomfort dimension and 48% moderate problems on anxiety/depresion dimension. Were no extreme issues on mobility dimension 21.2%, self care 2.8%, usual activities 17.3%, pain/discomfort 35.7% and anxiety/depression 19.7% [2].

Table 5. Health status of DM outpatient at Karanganyar Public Hospital

EQ-5D-5L Profiles	No Problem	Slight Problems	Moderat problems	Severe problems	Extreme problems
Mobility	61 %	34 %	5 %	-	-
Self care	89 %	11 %	-	-	-
Usual activities	39 %	54 %	7 %	-	-
Pain/Discomfort	18 %	37 %	41 %	4 %	-
Anxiety/Depression	17 %	35 %	48 %	-	-

VAS

VAS Measurement was aimed to see description of DM outpatients health status in Karanganyar public hospital, patients being asked to cross the scale from 0 to 100 on the questionnaire. A value of 0 indicates poor patient health, and a value of 100 indicates a healthy patient, resulted average of DM outpatients health status in Karanganyar public hospital was 80.1, it means DM outpatients in Karanganyar Public Hospital included on healthy category, while average VAS score was 74.3 [2].

Utility Index

Assessment of utility value based on Thailand value set, resulted average DM outpatient utility value was 0.833 while average utility value was 0.870 [2].

Table 6 shown DM outpatient's characteristic on this study were male patients are better average utility value (0.853) than female, same result with male utility value (0.826) better than female [9].

Mann-Whitney statistic test result of gender p=0.034 there was significant difference on utility value ($p<0.05$), it may concluded that gender influential about the value of utility. DM outpatient who ages >60 has lowest utility value (0.795) than younger patient. Statistical results test p=0.001 there was significant difference of utility value ($p<0.05$) hence concluded age can influence utility value. The life quality decreased along with aging [10]. Lowest utility value of education level was patient who has low education level (0.800). Statistical results test p=0.001 there was significant difference of utility value ($p<0.05$) hence conluded education influential about the value of utility. Same result with were higher level education has better quality of life [11]. Average utility value of DM outpatients who were unemployed was 0.825. The result of statistical test (p=0.151) there was no significant difference of utility value ($p>0.05$) may be concluded that job has no relation to utility. Same result with there were no significant affect between job and DM outpatient utility value however unemployed outpatient doesn't mean has no tough daily activity, for the example housewife does several tough activity such as sweeping, mopping, washing, and cooking [12]. Average utility value of DM outpatient who has length of DM disease more than 5 years was 0.793 it was lower than patient who has suffering less than 5 years. Statistical test results p=0.000 there was a significant difference ($p<0.05$) period of sickness may affect utility value. Patient who has been suffer for so long and having high blood glucose has big possibility to have a nerve damage or neuropathy type 2 diabetics.

Table 6. DM outpatient's characteristic and utility value

Characteristic	n	Average	*P*
Gender			
Male	56	0.853 ± 0.092	0.034
Female	44	0.807 ± 0.108	
Age			
18-40 years old	13	0.893 ± 0.065	0.001
41-60 years old	41	0.856 ± 0.100	
>60 years old	46	0.795 ± 0.102	
Education level			0.001
Low (Elementary School)	50	0.800 ± 0.107	
Medium (Senior High School)	29	0.841 ± 0.080	
High (University)	21	0.900 ± 0.081	
Job Status			0.151
Employed	24	0.858 ± 0.083	
Unemployed	76	0.825 ± 0.106	
Length of DM disease			0.000
<5 years	47	0.878 ± 0.080	
>5 years	53	0.793 ± 0.103	

Note :
Mann-Whitney Test
Kruskal-Wallis Test

Result of therapy pattern and utility value for single oral antidiabetic therapy was 0.855 means higher quality of life where metformin as single antidiabetic used, same result from Dinaryanti's study declared that quality of life in diabetic outpatients who use metformin were better than other antidiabetic because metformin is the most effective achieving fasting blood sugar and post prandial levels [5]. Result of Kruskal-Wallis test (p=0.004) has significant relation ($p<0.05$) it may be concluded therapeutic pattern has influence with utility, shown in Table 7.

Table 7. Therapy pattern and utility value

Therapy Pattern	n	Average	*P*
Single oral antidiabetic	29	0.885 ± 0.086	0.004
Combination oral antidiabetic	27	0.820 ± 0.094	
Single insulin antidiabetic	17	0.848 ± 0.090	
Combination insulin antidiabetic	7	0.783 ± 0.138	
Oral insulin antidiabetic	20	0.779 ± 0.098	

Table 8 shown average utility value of uncontrolable blood glucose level was 0.802 lower than controlable glucose level, result of Mann-Whitney test (p=0.002) there was significantly difference of utility value (p<0.05), hence concluded glucose level influence to utility. Utility value of DM patient who has controlable blood glucose level was higher than uncontrolable patient [13].

Table 8. Blood glucose level and utility value

Blood Glucose Level	n	Average	*p*
Controlable	46	0.869 ± 0.080	0.002
Uncontrolable	54	0.802 ± 0.108	

Table 9 shown average utility value of DM outpatient with complication was 0.754 while patient without complication achieve greater result, Mann-Whitney test result p=0.000 there was significantly difference of utility value (p<0.05), hence concluded complication effect to utility value same with who has significant correlation between complication to patient's quality of life [14]. Prospective clinical test results from the United Kingdom Prospective Diabetic Study and Diabetic Control and Complication Trial have proven that improving blood glucose control may reduce the complications of retinopathy, nephropathy, neuropathy and cardiovascular disease in diabetes mellitus [15].

Table 9. Compications and utility value

Complications	n	Average	*p*
Complication	32	0.754 ± 0.094	0.000
Non complication	68	0.870 ± 0.083	

Study limitations

Limitation of study with cross sectional design has a weakness may not know the patient's medical history and the effects of antidiabetic use, patients quantity based on variable therapy pattern, unproportional complications and amount of samples in this study is limited due to policy of the hospital regarding patient data.

4. Conclusion

Average direct medical cost of DM outpatient at Karanganyar Public Hospital was IDR 433,728.00. Biggest component cost was pharmacy. Therapy pattern, blood glucose level and complication has significant correlation to direct medical cost. Average DM outpatient utility value was 0.833. DM outpatient's characteristic, therapy pattern, blood glucose level and complication has significant impact to utility value.

Hospital need to consider the analysis of the medicine therapy cost in determining the medicine therapy which will be given to the patients and hospital need to maintain and improve the service quality and gave education to DM outpatients in order to maintain and increasing the utility.

References

[1] Wells B G, Dipiro J T, Schwinghammer T L and Dipiro C V 2015 9th *Pharmacotherapy Handbook* New York 161.

[2] Kemenkes RI 2013 *Guidance in the Impementation of Pharmacoeconomic Studies* Jakarta 24.
[3] Sakamaki H, Ikeda S, Ikegami N, Uchigata Y, Iwamoto Y, Origasa H, Otani T and Otani Y 2006 Measurement of HRQL using EQ-5D in patients with type 2 diabetes melitus in Japan *Value in Health* **9** 47-53.
[4] Hartanto D and Mulyani T 2017 The description of the diabetes mellitus type 2 patient cost with oral antidiabetic therapy at Ulin Public Hospital in Banjarmasin *Scientific Journal of Ibnu Sina* **2** (1) 114.
[5] Dinaryanti P, Fudholi A and Andayani T M 2012 Cost and effectiveness analysis of type 2 diabetes mellitus outpatient treatment at RSUD Sleman Yogyakarta *Journal of Management and Pharmaceutical Care* **5**(1) 19.
[6] Andayani T 2006 Analysis of diabetes mellitus therapy cost at dr. Sardjito public hospital in Yogyakarta Indonesia *Pharmaceutical Journal* **17**(3) 130-135.
[7] Baroroh F, Solikah W Y and Urfiyya Q 2016 Analysis of diabetes mellitus type 2 treatment cost at PKU Muhammadiyah Hospital in Bantul Yogyakarta *Journal of Pharmaceutical Science and Practical* **1**(2) 15-17.
[8] Ndraha S 2014 *Diabetes Melitus Type 2 and the Current Governance Internal Disease* Department of Medical Faculty of Krida Wacana University Jakarta **27**(2) 16.
[9] Mateo D C, Gordillo M A G, Oliveres P R and Adsuar J C 2015 Normative values of EQ-5D-5L for diabetes patients from Spain *Nutr. Hosp.* **32**(4) 1595-1602.
[10] Golicki D, Dudzinska M, Zwolak A and Tarach J S 2015 Quality of life in patients with type 2 diabetes in Poland – comparison with the general population using the EQ-5D Questionnaire *Original Papers* **24**(1) 139-146.
[11] Saleh F, Ara F, Mumu S J, and Hafez 2015 Assessment of health-related quality of life of Bangladesh patients with type 2 diabetes using the EQ-5D: a Cross-sectional study *BMC Res. Notes* (**8**) 497.
[12] Trisnawati S K and Setyorogo S 2013 The Factor of the risk of diabetes melitus type 2 case at Puskesmas in Cengkareng Sub-district West of Jakarta on 2012 *Health Scientific Journal* **5**(1) 8-10.
[13] Zurdayanis and Marfilanti E 2010 The correlation of blood glucose level with the quality of life the patient of diabetes mellitus type II at RSUD Sleman Yogyakarta *Universitas Islam Indonesia* **2** (6) 34.
[14] Ningtyas D W, Wahyudi P and Prasetyowati I 2013 The analysis of quality of life of patient with diabetes melitus type ii at Bangil Public Hospital of Pasuruan District *Scientific Article of College Students Research Result* p 6 [5 Mei 2017].
[15] Issa B A and Baiyewu O 2006 Quality of life patients with diabetes mellitus in a Nigerian Teaching Hospital *Hong Kong J Psychiatry* **16** 27-33.

Validation of behaviour measurement instrument of patients with diabetes mellitus and hypertension

G Z Saputri, Akrom, S M Dini

Faculty of Pharmacy, Universitas Ahmad Dahlan, Yogyakarta, Indonesia

E-mail: zukhruf.alparslan@gmail.com

Abstract. Non-adherence to the treatment of chronic diseases such as hypertension and Diabetes Mellitus (DM) is a major obstacle in achieving patient therapy targets and quality of life of patients. A comprehensive approach involving pharmacists counselling has shown influences on changes in health behaviour and patient compliance. Behaviour changes in patients are one of the parameters to assess the effectiveness of counselling and education by pharmacists. Therefore, it is necessary to develop questionnaires of behaviour change measurement in DM-hypertension patients. This study aims to develop a measurement instrument in the form of questionnaires in assessing the behaviour change of DM-hypertension patients. Preparation of question items from the questionnaire research instrument refers to some guidelines and previous research references. Test of questionnaire instrument valid was done with expert validation, followed by pilot testing on 10 healthy respondents, and 10 DM-hypertension patients included in the inclusion criteria. Furthermore, field validation test was conducted on 37 patients who had undergone outpatient care at the PKU Muhammadiyah Yogyakarta City Hospital and The Gading Clinic in Yogyakarta. The inclusion criteria were male and female patients, aged 18-65, diagnosed with type 2 diabetes with hypertension who received oral antidiabetic drugs and antihypertensives, and who were not illiterate and co-operative. The data were collected by questionnaire interviews by a standardized pharmacist. The result of validation test using Person correlation shows the value of 0.33. The results of the questionnaire validation test on 37 patients showed 5 items of invalid questions with the value of r <0.33, e: questions 2, 3, 6, 10 and 11, while the other 10 questions show the value of Pearson correlation > 0.33. The reliability value is shown from the Cronbach's alpha value of 0.722 (> 0.6), implying that the questionnaire is reliable for DM-hypertension patients. This Behavioural change questionnaire can be used on DM-hypertension patients, and an FGD approach is required for the development of factors affecting this questionnaire.

Keywords: DM-hypertension, behaviour, compliance, validation, reliability

1. Introduction

Diabetes mellitus (DM) is a chronic condition that can cause many serious complications, such as cardiovascular, nephropathic, retinopathic, and neuropathic diseases, which raise chronic illness and mortality rates [1]. Data show that every seven minutes a person succumbs to diabetes mellitus [2]. According to WHO data, it is estimated that the world population will undergo an increase in diabetes mellitus prevalence in the period of 2000-2030 from 37% to 114% [3]. The International Diabetes

Federation also predicts that 246 million adults have DM. DM prevalence in adults in the world has reached 6.4% and will rise to 7.7% in 2030. It is projected that between 2010 and 2030 there will be a 69% increment of adult DM patients in developing countries, and 20% in developed ones [4].

High blood pressure is a significant predictor of nephropathy, retinopathy, and cardiovascular disorder in type 2 diabetes [4]. Prior studies emphasised the importance of diabetes education from professional health services on DM therapy and management, lifestyle education including diet, physical exercise, blood sugar level monitoring, foot treatment, and treatment modification that can improve the clinical outcome and quality of life of DM patients [5, 6, 7]. Educated knowledge about DM management influences changes in the behaviour and self-management of DM patients [7, 8].

Three domains that affect behaviour are cognitive, affective and psychomotor [9, 10]. Behaviour is one's response or reaction to a stimulus. Behaviour is determined by three main factors: Predisposing factors, or factors that precede behaviour, which account for the reason and motivation for certain behaviour, such as knowledge, beliefs, attitude and demography; enabling factors, such as the availability and affordability of health resources and skills related to health, which allow particular behaviour to materialise; reinforcing factors, which accompany behaviour and play a role in establishing certain behaviour, including family, peers, teachers, and health workers [11, 12, 13].

Previous research on the relationship between knowledge about DM, health beliefs, and DM management revealed that the knowledge of DM patients augments their comprehension of DM management (β = 0.262, t = 3.328, p = 0.001), and that health beliefs affect DM patients' knowledge (β=0.07865, t=2.439, p<.016). Such investigation employed the "Health Belief Model" theoretical approach to explain self-care activities, such as recommended DM management, and focus on behaviour relating to disease prevention [7, 14].

The American Association of Diabetes Educators has developed standards for measuring the outcomes of DM self-management education. One of the recommended standards of diabetes self-care behaviour in DM management is assessing seven self-care behaviours, comprising physical activities, diets, the use of medicinal therapy and management, blood sugar level monitoring, overcoming hypoglycaemia occurrences, declines in DM complications, and psychosocial adaptation in living life [7,8].

A preceding study found that education and counselling by Brief Counselling 5A on hypertension patients can influence changes in their behaviour [9]. Modified Brief Counselling 5A is developed through five stages: Assess, the stage of current behaviour, the importance of changing it, the level of readiness for change, and social support; Advise, where individual behaviour clearly and specifically changes, including information about personal health gains and losses; Agree, which is the stage of collaboration towards the targeted behaviour; Assist, where help is given to solve various problems in order to achieve the expected behavioural change; Arrange, in which further actions are taken to evaluate progress and hurdles, and to adjust planning [15].

The behavioural change of a patient into greater compliance can be used as parameter in assessing the effectiveness of education and counselling by a pharmacist [2]. Instruments measuring therapeutic compliance have been widely developed, but instruments gauging patients' behaviour by assessing their change and readiness have yet to receive the same attention, raising the need to develop a DM-hypertension patient behaviour measurement instrument.

2. Materials and Methods

2.1. Methods

This study was reviewed and approved by the Research Ethics Committee of Universitas Ahmad Dahlan, numbered EC 011610143. The arrangement of question items in the research questionnaire instrument referred to guidelines and prior studies. The development of the DM-hypertension patient behaviour measurement instrument utilised a research and development design by a pharmacist and a researcher. The questionnaire draft was then reviewed by a psychology expert and an expert researcher of endocrines. The validity test for the questionnaire instrument was done by expert validation, with pilot testing on 10 healthy respondents and 10 DM-hypertension patients within the inclusion criteria

[2]. Next, a field validation test was conducted on 37 patients under outpatient care at the PKU Muhammadiyah Hospital and the Gading Clinic, both in the city of Yogyakarta [11, 12]. The inclusion criteria were male and female patients aged 18-65 diagnosed with type 2 diabetes with hypertension who received oral antidiabetic drugs and antihypertensives, and who were not illiterate and co-operative. Data were collected by means of questionnaire interviews given by standardised pharmacists. Statistical analysis of validity used Pearson correlation, and factor analysis. Reliability test used Cronbach's alpha value.

2.2. Materials

The DM-hypertension patient behaviour measurement instrument consists of 15 question items in cognitive, affective and psychomotor domains, given in Indonesian. The cognitive domain contains questions on knowledge about DM-hypertension disorders, complications, and the therapeutic management of oral antidiabetic and antihypertensive drugs. The affective domain comprises questions related to the feelings and preconceptions of patients regarding regular oral antidiabetic and antihypertension medications to control their blood sugar and blood pressure levels. The psychomotor domain enquires patients' behaviour towards compliance in taking medications, diet including the consumption of carbohydrate, sugar and salt, and ceasing medication without consulting the doctor.

The questionnaire employs yes-no questions, for which a "yes" answer is scored 1 and a "no" is given 0, except for items number 9, 14, and 15 where a "no" is scored 1, and a "yes" response 0. Patient behavioural levels measured by the questionnaire are categorised into pre-contemplation (if there is no affirmative answer in the cognitive, affective, and psychomotor domains), contemplation (if the answers are affirmative only in the cognitive domain), preparation (if the answers are affirmative only in the cognitive and affective domains), and action (if the answers are affirmative in all three domains).

3. Results and Discussion

This research aims to develop a DM-hypertension patient behaviour measurement instrument. The arrangement of the question items referred to previous studies and DM-hypertension management guidelines. The questionnaire validation test was carried out at the PKU Muhammadiyah Yogyakarta City Hospital and the Gading Clinic in Yogyakarta from June to August 2016. As subjects of this preliminary study, 45 DM type 2 patients with hypertension meeting the inclusion criteria were obtained, but 8 patients were excluded because they could not follow the research to the end, leaving 37 patients for this validation test. Description of the demographical characteristics of patients in this study is presented in Table 1.

The patients were predominantly women (70.3%), married (91.9%), and aged 60 or over (59.5%), and their last education was mostly at senior high school (48.6%). Most of them were merchants, pensioners and housewives (62.2%), and were covered by the government (BPJS) (78.4%). The majority of patients did not have diabetes mellitus history (51.4%), practiced a diet controlling the consumption of salt, sugar and coffee (89.2%), and did not smoking (94.6%).

Table 1. Demographical characteristics of patients for questionnaire validation

Demographical characteristic	n	No (%) / Mean ±SD
Sex		
Female	26	70.3%
Male	11	29.7%
Marital status		
Married	34	91.9%
Not married	3	8.1%
Age		
<60	15	40.5%
≥60	22	59.5%
Body weight (kg)		64.62 ±11.82
Body height (m)		1.57 ±0.07
Body mass index		25.97 ±4.10
Education		
Primary school	9	24.3%
Junior high school	6	16.2%
Senior high school	18	48.6%
D1/D2/D3/S1	3	8.1%
No school education	1	2.7%
Occupation		
Civil servant	1	2.7%
Entrepreneur	4	10.8%
Private employee	3	8.1%
Unemployed	5	13.5%
Other	23	62.2%
Payment		
Self-funded	2	5.4%
BPJS	29	78.4%
Other insurance	6	16.2%
Diabetes mellitus history		
Yes	18	48.6%
No	19	51.4%
Diet		
Salt and sugar	3	8.1%
Coffee	1	2.7%
Salt, sugar and coffee	33	89.2%
Habit		
Smoking	2	5.4%
Not smoking	35	94.6%

3.1. Behavioural Level Questionnaire Validation

Results of the behavioural level questionnaire validity and reliability tests are provided in Table 2. The validity test by Pearson correlation resulted in a value greater (>) than critical r, namely > 0.33. Validation of the questionnaire on 37 patients disclosed five invalid question items with $r < 0.33$, which were questions number 2, 3, 6, 10 and 11, while the other ten questions had a Pearson correlation coefficient of $r > 0.33$. Reliability was affirmed by a Cronbach's alpha value of 0.722 (>0.6), indicating that this behaviour measurement questionnaire is reliable for DM-hypertension patients.

The re-validity test was done by removing the invalid item, and a number of 10 question items showed the value of r > 0.3, with the Cronbach's alpha value being 0.722 (> 0.6) (Table 3). Several questions 2, 3, 6, 10 and 11 showed invalid, this can be seen from the answers of patients who are mostly uniform. This indicates that the questionnaire's differentiation is quite low, in addition to the question asked too easily so it can not distinguish the answers between patients. Question number 2 and 3 are designed in the cognitive domain, the whole patient answers "yes" and the uniform answer makes the validation value low. Become input for the researcher related to sentence question in item 2 and 3. While question number 6 is arranged in the affective domain, it also showed uniform answer so that validation value was low. In the psychomotor domain the behavioral adherence behavior is shown from items 10 and 11.

Table 2. Results of validation and reliability test of the DM-hypertension patient behavioural level questionnaire

No.	Question	Mean	SD	Pearson correlation (r>0.33) N=37	Cronbach's alpha (when item deleted)
COGNITIVE					
1	Do you know that diabetes mellitus with hypertension complication is a dangerous disease?	0.9189	0.27672	0.489	0.708
2	*Do you know that antidiabetic drugs must be used continuously to control blood sugar level?*			*(invalid)*	
3	*Do you know that antihypertension drugs must be used continuously to control blood pressure?*			*(invalid)*	
4	Do you know that untreated high blood pressure will cause blood vessel damage and trigger heart disorder and stroke?	0.9189	0.27672	0.651	0.697
5	Do you know that high blood sugar level and blood pressure will cause functional disorder of other organs, such as the kidneys and incurable wounds?	0.8919	0.31480	0.573	0.700
AFFECTIVE					
6	*Do you believe that using antidiabetic drugs can control blood sugar level?*			*(invalid)*	
7	Do you believe that using antihypertension drugs can control blood pressure?	0.9459	0.22924	0.459	0.712
8	Do you have the willingness to use antidiabetic and antihypertension drugs regularly to control blood sugar level and blood pressure?	0.9730	0.16440	0.456	0.717
9	Do you presume that using antihypertension drugs continuously can damage the kidneys?	0.4324	0.50225	0.373	0.719
PSYCHOMOTOR					
10	*Do you always bring antidiabetic drugs when travelling or working?*			*0.058 (invalid)*	
11	*Do you always bring antihypertension drugs when travelling or working?*			*0.134 (invalid)*	
12	Have you reduced sugar consumption to control blood sugar level?	0.9459	0.22924	0.329	0.719
13	Have you reduced salt consumption to control blood pressure?	0.8108	0.39706	0.345	0.716
14	Will you stop using antidiabetic drugs if the clinical symptoms of diabetes mellitus do not occur?	0.6486	0.48398	0.750	0.666
15	Will you stop using antihypertension drugs if the clinical symptoms of hypertension do not occur?	0.6757	0.47458	0.796	0.660

Table 3. Result of re-validity test of the DM-hypertension patient behavioural level questionnaire

No	Item of Question	r (Pearson correlation)	Validity
Cognitive			
1	Item 1	0.489	Valid
2	Item 4	0.651	Valid
3	Item 5	0.573	Valid
Affective			
4	Item 7	0.459	Valid
5	Item 8	0.456	Valid
6	Item 9	0.373	Valid
Psychomotor			
7	Item 12	0.329	Valid
8	Item 13	0.345	Valid
9	Item 14	0.750	Valid
10	Item 15	0.796	Valid

Table 4. Percentage distribution of the respondents' answers to each question item of the DM-hypertension patient behavioural level questionnaire

No	Question	Yes	No
COGNITIVE		N (%)	N (%)
1	DM-hypertension is a dangerous disease.	34 (91.9)	3 (8.1)
2	*DM drugs are used regularly to control blood sugar level.*	*37 (100)*	*0 (0)*
3	*Antihypertension drugs are used regularly to control blood pressure.*	*37 (100)*	*0 (0)*
4	Untreated high blood pressure will trigger heart disorder and stroke.	34 (91.9)	3 (8.1)
5	High blood sugar level and blood pressure will cause functional disorder of other organs, such as the kidneys and incurable wounds.	33 (89.2)	4 (10.8)
AFFECTIVE			
6	*Belief in the regular use of DM drugs to control blood sugar level*	*37 (100)*	*0 (0)*
7	Belief in the regular use of antihypertension drugs to control blood pressure	35 (94.6)	2 (5.4)
8	Willingness to use antidiabetic and antihypertension drugs regularly to control blood sugar level and blood pressure	36 (97.3)	1 (2.7)
9	Presumption that long-term drug use damages the kidneys	16 (43.2)	21 (56.8)
PSYCHOMOTOR			
10	*Always carrying DM drugs when travelling*	*34 (91.9)*	*3 (8.1)*
11	*Always carrying antihypertension drugs when travelling*	*35 (94.6)*	*2 (5.4)*
12	Sugar diet to control blood sugar level	35 (94.6)	2 (5.4)
13	Salt diet to control hypertension	30 (81.1)	7 (18.9)
14	Ceasing DM medication when clinical symptoms abate	24 (64.9)	13 (35.1)
15	Ceasing antihypertension medication when clinical symptoms abate	25 (67.6)	12 (32.4)

The results of the validation test items 10 and 11 are very low, this is because the repetition of almost identical questions turns the answer "yes" to the respondents. Most of the answers are uniform (yes = 91-94%), so the power difference is quite low. The invalidity of questions number 2, 3, 6, 10 and 11 can be observed from the virtually uniform responses given by patients. This implies that the differentiating power of the questionnaire was rather low, coupled with overly easy questions that could not distinguish responses among patients. Further research is needed to develop this instrument through focus group discussions (FGD) to explore the understanding levels of patients. The percentage distribution of the respondents' answers to each question item are given in Table 4.

Ten valid items were then tested for validity by factor analysis to see the correlation among fellow indicators simultaneously [18]. The Determinant of Correlation Matrix value is 0.001, this value approaches 0 thus the correlation matrix between the variables is interrelated. The results showed that the value of Kaiser Meyer Olkin (KMO) Measure of Sampling of 0.521. The KMO complied requirement because the value above 0.5. The calculation results with SPSS resulted in a Barlett Test of Spehricity of 210,643 with a significance of 0.000. Indicated that Bartlett Test of Spehricity fulfilled the requirements because of significance below 0.05 (<5%). While the matrix component of this questionnaire were 4 factor components, with each component matrix value or factor loading > 0.5. This indicated that ten question items were valid for use in patients with DM-hypertension [18].

Based on the Trans-Theoretical Model (TTM) theory developed by DiClemente, patient behaviours are classified into pre-contemplation, contemplation, preparation, action and maintenance [9, 14]. The distribution of the patients' behavioural levels is displayed by Table 5. TTM was designed to acquire direct impact (expected/deliberate behavioural change) on individuals or groups. This model is the most widely used concept of behavioural change in health behaviour studies [16, 17].

Table 5. Distribution of DM-hypertension patient behavioural level

Behaviour category	N (%)
Pre-contemplation	6 (16.2)
Contemplation	17 (45.9)
Preparation	7 (18.9)
Action	7 (18.9)

At the pre-contemplation stage, the individual does not consider or intend to alter problematic behaviour (start healthy behaviour) soon (within six months), because the patient usually does not understand facts concerning risks posed by unhealthy behaviour. Such patients need encouragement to become more aware of their decisions and the benefits of adopting healthy behaviour. The contemplation stage is where the patient becomes aware and willing to change problematic behaviour, and is able to consider the advantages and disadvantages of changes that one will face. The preparation stage sees the individual ready to take action in the near future (in 30 days ahead). At this level, the gains (pros) which support change towards healthy behaviour are greater than the hurdles (cons), and patients take small steps that they believe can help make healthy behaviour part of their lives. The action stage is where patients have altered their behaviour in the last six months and work hard to keep going forward. In TTM, an individual attaining this phase has undergone half the changing process, and thus is most likely to undergo relapses to unhealthy behaviour. If one is not sufficiently prepared for change and committed to planning actions of their choice, reversions to problematic behaviour would be largely possible. At the maintenance stage, the patient has successfully achieved and maintained behavioural change for at least six months. Risks of setbacks are still present but not too considerable, and the individual tries to keep practicing healthy behaviour [9, 14, 15, 17].

This inquiry found that most of the patients were at the contemplation behavioural level (45.9%). This suggests that the majority of patients had the awareness and will to change their behaviour to action or maintenance level. However, re-monitoring is required at least in second visits as follow-up to this research. Nonetheless, 16.2% patients were at pre-contemplation level, indicating that they were not yet prepared for and had not cognitively comprehend knowledge about DM-hypertension disease, and had not had the disposition to take medications regularly. This calls for intervention in the shape of education regarding DM-hypertension management and the motivation to improve patient compliance.

Limitations in this study are still needed FGD in digging factors that affect the understanding of patients, and compilation of sentences in the questionnaire items questionnaire. Interventions are needed in the form of education related to DM-Hypertension management and motivation in improving patient compliance.

4. Conclusions
This research showed that the DM-hypertension patient behaviour measurement questionnaire was valid and reliable. Further research in the form of FGDs is required to divulge factors which affect patients' understanding and responses, and to evaluate the impact of intervention on their behavioural changes.

Acknowledgement
Thanks to all DM-hypertension patient respondents for their participation; hopefully, this study will be useful for them. Thanks to PKU Muhammadiyah Yogyakarta City Hospital and the Gading Clinic for permitting this research in their premises, to all healthcare staff including the nurses, doctors and pharmacists serving at PKU Muhammadiyah Yogyakarta City Hospital and the Gading Clinic, and to the pharmaceutical counsellors who assisted in collecting data, Ainun Muthoharoh, S.Farm., Apt., Sari Rahmadhani, S.Farm., Apt., Gilang, S.Farm., Apt., Sri, S.Farm., Apt., along with the research team.

References
[1] Shrivastava S R, Shrivastava P S, Ramasamy J 2013 Role of self-care in management of diabetes mellitus *J. Diabetes Metab. Disord.* Mar **5** 12-14.
[2] Lai P S, Chua S S, Tan C H and Chan S P 2012 Validation of the diabetes, hypertension and hyperlipidemia (DHL) knowledge instrument in Malaysia *BMC Med. Res. Methodol.* **12**(1) 18.
[3] Sriram S, Chack L E, Ramasamy R, Ghasemi A, Ravi T K, Sabzghabaee A M 2011 Impact of pharmaceutical care on quality of life in patients with type 2 diabetes mellitus *J. Res. Med .Sci* **16**(Suppl1) S412–8
[4] Chobanian A V, Bakris G L, Black H R, Cushman W L, Green l A, Izzo J l, Jones D W, Materson B J, Oparil S, and Wright J T 2003 *JNC VII Express: the seventh report of the Joint National Committee on prevention, detection, evalution and treatment of high blood presure* U.S. Deparment of Health and Human Services 12-33
[5] Colleran K M, Starr B, Burge M R 2003 Putting diabetes to the test: Analyzing glycemic control based on patients' diabetes knowledge *Diabetes Care* **26**(7) 2220-2221
[6] McPherson M L, Smith S W, Powers A, Zuckerman I H 2008 Association between diabetes patients' knowledge about medications and their blood glucose control *Res. Social. Adm. Pharm.* **4**(1) 37-45.
[7] Samuel O A 2014 Diabetes knowledge, health belief, and diabetes management among the Igala Nigeria *SAGE Open* April-June 2014 1–8.
[8] American Diabetes Association 2004 Diagnosis and classification of diabetes mellitus *Diabetes Care* **27**(Suppl 1) S5-S10
[9] DiClemente C, Delahanty J, Jean F, Earley M, Garay M, Preston G, Meredith H, Angela P, Kristina S, Onna V, and Katherine W 1994 Health and addictive behaviours investigating transtheoritical solution (online) *Transtheoritical Model of Behaviour Change Measure*
[10] Bloom B S (Ed.). Engelhart M D, Furst E J, Hill W H, Krathwohl D R 1956 *Taxonomy of Educational Objectives, Handbook I: The Cognitive Domain* (New York: David McKay Co Inc).
[11] Saputri G Z, Akrom A, Darmawan D 2016 Counseling and motivational short text messages increase adherence and behavioral changes in patient with hypertension *Jurnal Kedokteran and Kesehatan Indonesia* **7**(3) 87–94.
[12] Notoatmodjo S 2010 *Promosi Kesehatan and Perilaku Kesehatan* (Jakarta: Rineka Cipta)

[13] Notoatmodjo S 2012 *Promosi Kesehatan and Perilaku Kesehatan* (Jakarta: Rineka Cipta) 137-147

[14] Becker M H and Maiman L A 1975 Socio-behavioural determinants of compliance with health and medical care recommendations *Medical Care* **13** 10–24.

[15] Lenio J A 2006 Analysis of the Transtheoretical model of behavior change *Journal of Student Research University of Wisconsin-Stout* 5th edn 73–86.

[16] Jake M, Mariella M, Norman D, Liz O'Brein 2012 *Forestry, sustainable 1 behaviours and behaviour change: Theories.*

[17] Glanz K, B K Rimer, and K Viswanath 2008 *Health behavior and health education: Theory, research, and practice* 4th ed. (San Francisco: Wiley).

[18] Plakas, Sotirios, Dimos M, Marianna M, Theodoula A, Georgia F 2016 Validation of the 8-item morisky medication adherence scale in chronically ill ambulatory patients in rural *Greece Open Journal of Nursing* **6** 158-169.

IOP Conf. Series: Materials Science and Engineering **259** (2017) 012015 doi:10.1088/1757-899X/259/1/012015

Relation of knowledge and level of education to the rationality of self-medication on childhood diarrhea on the Code River banks in Jogoyudan, Jetis, Yogyakarta

H Dania, M N Ihsan

Faculty of Pharmacy, Universitas Ahmad Dahlan, Yogyakarta, Indonesia

E-mail: fizadan.djogja@gmail.com

Abstract. Self-medication as an alternative is used to reduce the severity of diarrhea. Optimal treatment can be done by increasing the rationalization of self-medication on diarrhea. This can be achieved with good knowledge about self-medication, which is in turn influenced by level of education. The aim of this study was to determine the relationship of knowledge and education level to rationality of self-medication on childhood diarrhea around the Code River in Jogoyudan, Jetis, Yogyakarta. The study was conducted by cross-sectional analytical observational design. The subjects were mothers who had children aged 2-11 years who had experienced diarrhea and had self-medication. Questionnaires were used to assess the rationality of self-medication on children's diarrhea by the parents. The respondents were askeds to fill out about indications, right drugs, doses, time intervals and periods of drug administration. Data were analyzed using chi- square. It was showed that of 40 respondents, 14 respondents (35%) performed rational self-medication on children's diarrhea and 26 respondents (65%). did not rationalize the treatment. The results of a bivariate test obtained a chi-square value of 9.808 (> 3.841) and a p value of 0.002 (<0.05) on the relationship between education level and rationality of self-medication and a chi-square value of 19.476 (> 3.841) and a p value of 0.000 (<0.05) on relationship between knowledge and rationality of self-medication. The conclusion of this study is that there is a correlation between knowledge and level of education and rationality of self-medication on childhood diarrhea on the Code River banks in Jogoyudan, Jetis, Yogyakarta.

Keywords: knowledge, education, rationalization, self-medication, diarrhea

1. Introduction

The Yogyakarta City Health Agency reported that diarrhea was one of the 10 most prevalent diseases in the city of Yogyakarta in 2015, comprising 10,774 cases. The reported data were based on the number of diarrhea patients who visited a Public Health Center (Puskesmas). The highest number of diarrhea cases was reported from Puskesmas Umbulharjo 1 with 1,354 cases, followed by Puskesmas Jetis with 1,149 cases and Puskesmas Tegalrejo with 1,128 cases [1]. One of the causative factors of diarrhea is polluted environment, such as river water contaminated by industrial or domestic waste [2]. Diarrhea in children can prompt dehydration that could lead to mortality. An effort that can be made to overcome severity and prevent death by dehydration due to diarrhea is by performing self-medication [3].

Self-medication should adhere to the terms of rational medication, so that optimal diarrhea treatment can be achieved and thus lower the risk of child mortality by diarrhea. The rationality of self-medication on diarrhea can be obtained with good knowledge about diarrhea, and the level of one's knowledge corresponds with the level of one's education. The higher the education, the more readily a person absorbs information, and consequently the more knowledge is gained [4]. Therefore, this study was conducted to identify the relationship of knowledge and education level to the rationality of self-medication on child diarrhea on the banks of the Code River in Jogoyudan Kampong, Jetis Sub-district, Yogyakarta.

2. Materials and Methods

2.1. Materials

The data collection instrument for this research was a questionnaire, consisting of a series of questions divided into three sections: The first section enquired the respondents' personal details including name, sex, child's age, education level and income, the second section revolved around knowledge about diarrhea and self-medication against the disease, and the third examined the rationality of the diarrhea medication. Contents of the questionnaire covered:

a. The respondent's knowledge about diarrhea, including the definition of diarrhea, the dangers of diarrhea, and the treatment of diarrhea.
b. The respondent's knowledge about oral rehydration therapy (ORT), including how ORT is administered, how long it can be given, and how it is made.
c. The respondent's knowledge about ORT replacement, breastfeeding during diarrhea, and feeding during diarrhea.
d. The respondent's knowledge about diarrhea medications, namely New Diatab and Zinc.
e. The rationality of diarrhea medication, concerning the right drug, which is ORT to prevent dehydration and zinc to prevent diarrhea recurrences, and the right dose of ORT and zinc, including the right way and the right time to administer them.

As a measurement tool for research, the questionnaire had to be valid and reliable, and thus needed to undergo validity and reliability tests [5]. Respondents involved in the validity and reliability tests should have similar characteristics to those of the respondents at the main research site [6].

2.2. Methods

This study was reviewed and approved by the Research Ethics Committee of Universitas Ahmad Dahlan (KEP UAD), numbered EC 011612154. The composition of the question items in the research questionnaire instrument referred to several guidelines and previous studies conducted in 2013 around the landfill Banguntapan Yogyakarta. The research employed an analytical observational design with a cross-sectional approach. Preliminary study was conducted on 30 respondents to test the validity and reliability of the questionnaire. The inclusion criteria were mothers with a child aged 2 to 11 who had suffered from diarrhea and had been self-medicated, residing on the Code River banks in Jogoyudan Kampong, Gowongan Village Administration, Jetis Sub-district, and willing to fill out the questionnaire. The validity test utilized Pearson correlation by SPSS. Every question item was considered valid if the p value (the value at Sig. (2-tailed)) $< \alpha$, or if the counted r was greater than the critical r. The number of 30 respondents has a critical r of 0.361; the instrument was hence deemed valid if the counted $r \geq 0.361$. Reliability was identified with a Cronbach's alpha analysis to determine the coefficient of reliability and to then verify whether the instrument was highly reliable. An instrument is seen reliable if it has a Cronbach's alpha greater than 0.6 [7]. A valid and reliable questionnaire would accurately gauge the level of knowledge and rationality of self-medication on children.

Each answer to the knowledge questionnaire was assigned a score; every correct answer was scored 1 and every wrong answer 0. The scores were summed up, then divided by the number of question items and finally multiplied by 100%, resulting in percentage. The respondent's knowledge fell into low category if the percent score was less than the average percentage, and entered high category if the percent score was equal to or greater than the mean percentage. This classification was based on the

IOP Conf. Series: Materials Science and Engineering **259** (2017) 012015 doi:10.1088/1757-899X/259/1/012015

total average value of all the respondents' scores. The self-medication was rational if all the question items in the rationality section of the questionnaire were correctly answered by the respondent.

The acquired data were analyzed by:

1. Descriptive analysis, to identify the characteristics of the research subjects, including the community group (RW) of the respondents' residence, their education level, their knowledge level and the rationality of their self-medication.
2. Bivariate statistics, to establish the relationship of knowledge and education level to the rationality of self-medication on child diarrhea by chi-square analysis.

The hypothesis to identify the correlation of knowledge and education level with the rationality of self-medication on childhood diarrhea utilized H_0. If the calculated chi-square value was greater than the critical value in the chi-square table, or the p value was less than 0.05, the H_0 was rejected and a relationship between knowledge and education level and the rationality of self-medication on children's diarrhea was established. If the counted chi-square was lower than the critical chi-square value, or the p value was over 0.05, the H_0 could not be rejected and thus no correlation between knowledge and education level and the rationality of self-medication on childhood diarrhea could be determined.

3. Results and Discussion

3.1. Questionnaire validity and reliability tests

Validity and reliability tests on the questionnaire was undertaken on 30 respondents outside the research sample (mothers with a child aged 2 to 11 who had had diarrhea and been self-medicated, residing in Jogoyudan Kampong RW 7, 8, and 9 which were not located on the Code riverbanks) with 30 question items, consisting of 20 items on knowledge about diarrhea and 10 items on the rationality of self-medication on childhood diarrhea. The validity test examined the correlation between the score of each item and the total score of all question items using Pearson correlation that was calculated with the SPSS software version 16. A question item was declared valid if its calculated r (r_{count}) was greater than the corresponding r from the critical value table (r_{table}). The r_{table} value of 30 respondents is 0.361 on a significance level of 5%. The validity test was performed on the sections on knowledge about diarrhea and on the rationality of self-medication on child diarrhea. Results of the validity test on the section on knowledge about diarrhea can be observed in Table 1.

Table 1. Validity test results of knowledge about diarrhea

No.	r_{count}	r_{table}	Note
P1	0.633	0.361	Valid
P2	0.076	0.361	Invalid
P3	0.697	0.361	Valid
P4	0.396	0.361	Valid
P5	0.162	0.361	Invalid
P6	0.396	0.361	Valid
P7	0.248	0.361	Invalid
P8	0.680	0.361	Valid
P9	0.404	0.361	Valid
P10	0.436	0.361	Valid
P11	0.342	0.361	Invalid
P12	0.507	0.361	Valid
P13	0.576	0.361	Valid
P14	0.526	0.361	Valid
P15	0.570	0.361	Valid
P16	0.322	0.361	Invalid
P17	0.443	0.361	Valid
P18	0.473	0.361	Valid
P19	0.542	0.361	Valid
P20	0.681	0.361	Valid

The validity test revealed that of the 10 question items in the section on knowledge about diarrhea, there were 3 items with an r_{count} of less than 0.361; consequently, those questions were dismissed and excluded from the research.

The reliability test for this survey was done by calculating the value of Cronbach's alpha. The questionnaire was considered reliable if the Cronbach's alpha of each variable was greater than 0.6. Results of the reliability test can be seen in Table 2 and Table 3.

Table 2. Reliability test result on knowledge about diarrhea

Variable	Cronbach's alpha	Note
Knowledge about diarrhea	0.827	Reliable

Table 3. Reliability test result on the rationality of self-medication on child diarrhea

Variable	Cronbach's alpha	Note
Rationality of self-medication on child diarrhea	0.635	Reliable

According to Table 2 and Table 3, the Cronbach's alpha values of the variables for knowledge about diarrhea and rationality of self-medication on childhood diarrhea were respectively 0.827 and 0.635, each of which was greater than 0.6. Therefore, all the items of the questionnaire were reliable and could be used for further studies.

3.2. Demographics characteristics

The number of respondents participating in this survey was 40 mothers who had children aged between 2 and 11 who had had diarrhea and been self-treated, living on the Code riverbanks in Jogoyudan Kampong, Jetis Sub-district, Yogyakarta. The disseminated questionnaires provided respondent data consisting of name, address/RW, and education level. The distribution of the addresses/RWs and education levels of the respondents is presented in Table 4 and Table 5.

Table 4. Distribution of the respondents' residential RW on the Code River banks in Jogoyudan Kampong, Yogyakarta

RW	Frequency	Percentage (%)
10	6	15
11	10	25
12	20	50
13	4	10
Total	40	100

Table 5. Distribution of the education level of respondents on the Code River banks in Jogoyudan Kampong, Yogyakarta

Education	Frequency	Percentage (%)
Low	22	55
High	18	45
Total	40	100

Education is one of factors that can affect the thoughts and actions of the respondents. Differences in their education level can cause differences in their assumed knowledge level as well, consequently influencing every action they take in managing self-medication on their children's diarrhea. This study defined education level as the last level of formal education undertaken by the respondents. Results showed that there were 22 respondents (55%) with low education levels or more than the 18 respondents (45%) with high education levels.

3.3. Questionnaire Results

3.3.1. The respondents' knowledge about diarrhea

Table 6. Distribution of the respondents' correct and incorrect answers per question item in the questionnaire on knowledge about diarrhea on the Code River banks in Jogoyudan Kampong, Yogyakarta

Question	Respondent Answer			
	Correct		Incorrect	
	Frequency	%	Frequency	%
P1	34	85	6	15
P2	22	55	18	45
P3	35	87.5	5	12.5
P4	40	100	0	0
P5	33	82.5	7	17.5
P6	12	30	28	70
P7	34	85	6	15
P8	15	37.5	25	62.5
P9	20	50	20	50
P10	27	67.5	13	32.5
P11	34	85	6	15
P12	17	42.5	23	57.5
P13	34	85	6	15
P14	13	32.5	27	67.5
P15	16	40	24	60

Question P1 was about the definition of diarrhea. The number of respondents who answered correctly was 34 (85%), while those who gave an incorrect answer were 6 respondents (15%). This indicates that there were more respondents who understood what diarrhea is than those who did not. Respondents who could not give an accurate answer to this question were probably less open to information concerning diarrhea, or were less certain about determining the frequency of children's liquid bowel movements which symptomizes diarrhea.

Question P2 asked about the possible consequences of diarrhea if not properly treated. There were slightly more respondents, 22 of them (55%), who chose the correct response than the 18 respondents (45%) who answered otherwise. These obtained data revealed that most of the respondents had already realized that diarrhea can be fatal if not handled well. The remainder were not aware that diarrhea can cause death because they thought that diarrhea was a mild disorder with no lethal potential.

Question P3 enquired about the first action that must be taken when a child contracts diarrhea. The table above shows that 35 respondents (87.5%) provided the exact answer, whereas those who responded incorrectly only comprised 5 respondents (12.5%). This means that the majority of respondents had acquired proper knowledge about the initial treatment that needs to be given when a child is infected with diarrhea, which is providing more drinking fluids to prevent dehydration that if not handled appropriately may lead to fatality.

Question P4 examined the administration of ORT to children with diarrhea. It transpired that all respondents (100%) answered correctly, implicating that all of them had already grasped that ORT is given to children when suffering from diarrhea.

Question P5 covered the respondents' knowledge about the duration of administering ORT to children with diarrhea. As many as 33 respondents (82.5%) selected the right answer, suggesting that most of the respondents knew that ORT should be given to diarrhea-suffering children until they are

IOP Conf. Series: Materials Science and Engineering **259** (2017) 012015 doi:10.1088/1757-899X/259/1/012015

relieved from the ailment. Conversely, 7 respondents (17.5%) picked the wrong answer, supposing that ORT is given only once at the first instance of diarrhea.

Question P6 concerned about giving ORT when the child vomits, in which case the administration should continue little by little but frequently. No more than 12 respondents (30%) had a correct response, much fewer than the 28 respondents (70%) with the erroneous one. This implies that many respondents had yet to grasp the method of administering ORT to children vomiting during diarrhea, with the presumption that forceful ORT treatment would exacerbate the vomiting.

Question P7 related to the making of ORT. In this matter, 34 respondents (85%) had already comprehended that ORT is made by dissolving the ingredients in boiled water, whilst 6 respondents (15%) had not.

Question P8 pertained to the alternative use of rice water as ORT substitute for diarrheic children. Most respondents, or 25 of them (62.5%), did not recognize that rice water can be given to ease symptoms and avert dehydration on children with diarrhea. This probably resulted from the respondents' different perception about rice water, as some of them did not understand precisely what was meant by rice water. Only 15 respondents (37.5%) were aware of this alternative medication.

Question P9 looked into the respondents' knowledge about breastfeeding diarrheic children, which should be maintained as intake to avoid the children from dehydration. This question was answered correctly and incorrectly by equally 20 respondents (50% each). Respondents who chose the wrong response assumed that breastfeeding could worsen the child's diarrhea.

Question P10 queried about knowledge relating to feeding children with diarrhea. A number of 27 respondents (67.5%) knew that children suffering from diarrhea should be fed more often than usual but with smaller portions, because in such condition their digestive system may not be able to sustain standard or larger portions.

Question P11 covered the knowledge of handling childhood diarrhea if blood is present in the stool. A child contracting bloody diarrhea should immediately be taken to the nearest health service center to receive further treatment for fear of diarrhea with complications. The majority of 34 respondents (85%) yielded the right answer, while the other 6 respondents (15%) did not because they thought that bloody diarrhea in children would subside by itself.

Question P12 asked about an antidiarrheal medicine named New Diatab. More specifically, the question assessed the knowledge level of respondents on the distinction between New Diatab and ORT. New Diatab possesses different properties from ORT, in that New Diatab contains kaolin and pectin which are useful in stopping diarrhea, whereas ORT contains substances that help recover body fluids lost during diarrhea to prevent dehydration. The number of respondents who understood the different properties of New Diatab and ORT was 17 (42.5%), fewer than the 23 respondents (57.5%) who did not know. This probably ensued from a lack of information gained by respondents regarding the effects of antidiarrheal drugs circulating in public.

Question P13 examined the knowledge that ORT can be bought at any pharmacy without a doctor's prescription. As many as 34 respondents (85%) were aware of this fact, more numerous than the 6 respondents (15%) who were not. The latter group was possibly less exposed to information about ORT and accustomed to receiving ORT only from Puskesmas when having their diarrheic children examined.

Question P14 and P15 were about the respondents' knowledge about zinc. Zinc is an antidiarrheal medication that averts recurrences of diarrhea in children, and is not an antibiotic, thus enabling purchase at pharmacies or drugstores without prescription. Question P14 measured respondents' understanding that zinc is not an antibiotic, which was realized by merely 13 respondents (32.5%). Question P15, meanwhile, gauged the respondents' knowledge about how to obtain zinc. Just 16 respondents (40%) were informed that zinc could be procured without prescription, fewer than the opposite group of 24 respondents (60%). These results uncovered that respondents' knowledge about zinc was still low, which could have stemmed from inadequate socialization to the public and the respondents' habit of acquiring zinc only from Puskesmas without buying it themselves.

From these acquired data, the distribution of respondents' level of knowledge about diarrhea was identified as described in Table 7.

Table 7. Distribution of respondents' level of knowledge about diarrhea on the Code River banks in Jogoyudan Kampong, Yogyakarta

Knowledge level	Frequency	Percentage (%)
Low	19	47.5
High	21	52.5
Total	40	100

The knowledge level will influence the measures taken in treating childhood diarrhea. The higher the knowledge level, the better the diarrhea treatment measures, which would bolster the realization of rational medication towards optimal treatment of diarrhea in children. The knowledge levels were divided into two categories, namely a low level when the respondent's average score was below the total average of all answers, and a high level when the respondent's average score surpassed the total average. Based on the questionnaires filled out by the respondents, results revealed that there were 19 respondents (47.5%) with a low knowledge level, fewer than those with a high knowledge level of 21 respondents (52.5%). The fact that the proportion of respondents with a low knowledge level was nearly half the total sample could have been instigated by uneven dissemination of information by relevant agencies around the Code riverbanks in Jogoyudan Kampong, so that many people there remained with low knowledge levels.

Nevertheless, these findings corroborate a prior study on 22 respondents in the vicinity of the landfill (TPA) of the Banguntapan Sub-district in Bantul, where 6 respondents (27.27%) had low knowledge and 16 respondents (72.73%) had high knowledge, i.e. the number of respondents with high knowledge exceeded that of the ones with low knowledge [8].

3. 3. 2. Rationality of respondents' self-medication on diarrheic children

Self-medication on childhood diarrhea can be optimal if the medication is rational. The rationality of self-medication includes the right indication, the right drug, the right dose, the right time interval for administration, and the right duration of medication. If any one of the five points is not met, the implemented medication is deemed irrational. Rationality could be assessed from the respondents' answers to the distributed questionnaires. Respondents fell into rational category if they correctly answered all the questions in the section on the rationality of self-medication on children's diarrhea.

Distribution of the respondents' correct and incorrect answers per question item questionnaire as can be seen in table 8.

Table 8. Distribution of the respondents' correct and incorrect answers per question item on the rationality of self-medication on child diarrhea on the Code River banks in Jogoyudan Kampong, Yogyakarta

Question	Respondent Answer			
	Correct		Incorrect	
	Frequency	%	Frequency	%
R1	33	82.5	7	17.5
R2	38	95	2	5
R3	32	80	8	20
R4	29	72.5	11	27.5
R5	24	60	16	40
R6	19	47.5	21	52.5
R7	26	65	14	35

Question R1 enquired about the right indication. The question cited a symptom indicating the occurrence of diarrhea on a child, which is liquid bowel movements taking place 4 times or more. If such indication emerges, diarrhea medication needs to be administered. As many as 33 respondents (82.5%) were able to respond correctly, whilst 7 respondents (17.5%) could not, denoting that the number of respondents who could identify the right indication predominated that of respondents who were not capable of doing so.

Question R2 related to the right drug. During diarrhea, children are given antidiarrheal medicine as management. The primary medication administered is ORT, which is effective in averting dehydration due to diarrhea, which may weaken the child and lead to death. Results disclosed that 38 respondents (95%) gave the right answer, selecting ORT to prevent dehydration caused by the child's diarrhea.

Question R3 queried about the time interval between ORT administrations. According to the "Buku Saku Lintas Diare" (Five Steps to Cure Diarrhea Pocketbook) published by the Ministry of Health of the Republic of Indonesia, ORT is given immediately after every event of defecation during diarrhea to avert dehydration. There were 32 respondents (80%) who answered the question correctly, by far more numerous than those who responded otherwise, implying that a large majority of the respondents had already applied the right time to administer ORT when their children suffered from diarrhea.

Questions R4 and R5 looked into the right dose of diarrhea medication for children. The proper dose of ORT to administer at the onset of diarrhea to prevent dehydration is 100-200ml after every instance of defecation. Aside from ORT, zinc is also required to prevent relapses of the child's diarrhea. Zinc should be given as much as one tablet or 20mg a day. Question R4 concerned the right dose of ORT. It transpired that 29 respondents (72.5%) had grasped the appropriate dose of ORT to give to children with diarrhea to avoid them from dehydration. This number was greater than that of respondents who did not use the right dose, consisting of 11 respondents (27.5%). Question R5 pertained to the right dose of zinc administration, to which 24 respondents provided the exact answer, implicating that they are capable of administering zinc with the right dose.

Question R6 asked about the right way to administer zinc. The right way closely relates to the right dose, and is an important aspect in which people frequently commit errors. The publicly available preparations of zinc tablets are zinc *dispertabs*, which are tablets that are not directly taken orally but must be dissolved with a tablespoon of boiled water at first. This is done to accelerate the drug's therapeutic effect and help children to ingest it. However, only 19 respondents (47.5%) knew the right way, fewer than the 21 respondents (52.5%) who did not. This could have stemmed from insufficient counselling, information and education provided by pharmacy staff when handing over the drugs, to the extent that respondents thought that zinc is used in a similar way to ordinary tablets.

Question R7 was about the right duration for giving ORT medication to diarrheic children. The 2011 edition of the Five Steps to Cure Diarrhea Pocketbook states that ORT should be administered until the patient is cured. Once the diarrhea is alleviated, the ORT treatment ceases. There were 26 respondents (65%) who selected the right answer for this question, more than the 14 respondents (35%) who took the wrong answer. It could thus be concluded that most of the respondents had already understood and been able to determine the time length for the medication of children suffering from diarrhea.

The gathered data yielded a distribution of respondents' rationality in the self-medication of diarrheic children as can be seen in Table 9.

Table 9. Distribution of the rationality of self-medication on child diarrhea on the Code River banks in Jogoyudan Kampong, Yogyakarta

Rationality	Frequency	Percentage (%)
Rational	14	35
Irrational	26	65
Total	40	100

The above table displays that those who could not answer all the questions correctly were 26 respondents (65%), more than the 14 respondents (35%) who were able to do so. These data point out that there were still many respondents who had not yet comprehended how to perform self-medication on diarrheic children in a rational way. Consequently, children's diarrhea may not be treated properly with optimal medication.

This finding contradicts results from previous research conducted around the TPA in Banguntapan Sub-district, Bantul, where 17 out of 22 respondents (77.27%) who practiced rational medication outnumbered those who performed irrational medication, consisting of 5 respondents (22.73%) [8]. This might have also resulted from the disproportionate distribution of information by related agencies in the area of the Code riverbanks in Jogoyudan Kampong. Another possible influential factor is the low education level of respondents, which made them less active and responsive towards seeking and gathering information concerning diarrhea.

3. 3. 3. Descriptive distribution of knowledge and education levels of respondents against the rationality of self-medication on diarrhea

The distribution of respondents' education level against the rationality of self-medication on child diarrhea is described in Table 10.

Table 10. Distribution of respondents' education level against the rationality of self-medication on child diarrhea on the Code River banks in Jogoyudan Kampong, Yogyakarta

Education	Rationality			
	Rational	%	Irrational	%
Low	3	21.4	19	73.1
High	11	78.6	7	26.9
Total	14	100	26	100

It is noticeable from Table 10 that respondents with low education level tended to act irrationally in practicing self-medication on childhood diarrhea, involving 19 respondents (73.1%), while merely 3 respondents (21.4%) could give rational self-medication to diarrheic children. Conversely, respondents with high education level were relatively able to perform self-medication on child diarrhea rationally, comprising 11 respondents (78.6%), a greater number than the 7 respondents (26.9%) who were not rational in applying self-medication on children's diarrhea. The collected data indicated that the higher the education level, the broader the assumed knowledge to the point of the capability to conduct rational self-medication on diarrheic children, attaining optimal medication. Meanwhile, the description of respondent knowledge against the rationality of self-medication on child diarrhea is provided in Table 11.

Table 11. Distribution of respondents' knowledge level against the rationality of self-medication on child diarrhea on the Code River banks in Jogoyudan Kampong, Yogyakarta

Knowledge	Rationality			
	Rational	%	Irrational	%
Low	0	0	19	73.1
High	14	100	7	26.9
Total	14	100	26	100

Table 11 shows that the knowledge level of respondents influences the rationality of their self-medication on child diarrhea. Good knowledge would support medication rationality, while deficient knowledge may make medication rationality not achieved. All 19 respondents with a low knowledge level gave irrational self-medication on childhood diarrhea, whereas among respondents with high

IOP Conf. Series: Materials Science and Engineering **259** (2017) 012015 doi:10.1088/1757-899X/259/1/012015

knowledge level, 14 were able to implement self-medication on diarrheic children rationally, and only 7 of them could not do so. Respondents with high knowledge level but performing irrational self-medication on children with diarrhea might have had knowledge limited to the disease and the names of antidiarrheal medicines, but had not enough knowledge about the right way to utilize those drugs.

The distribution of respondents' education level against their knowledge about childhood diarrhea is depicted in Table 12.

Table 12. Distribution of respondents' education level against knowledge about diarrhea on the Code River banks in Jogoyudan Kampong, Yogyakarta

Education	Knowledge			
	Low	%	High	%
Low	17	89.5	5	23.8
High	2	10.5	16	76.2
Total	19	100	21	100

One's education level is a factor that can affect the level of information gathered and knowledge learnt. The higher the education level, the more information is gained, leading to the acquisition of wider and better knowledge. This survey found that respondents with low education were inclined to have low knowledge as well. In contrast, respondents with a high education level tended to possess high knowledge accordingly. It can be observed from the above table that 17 of respondents with low education also had low knowledge, more than their 5 counterparts with high knowledge. On the other hand, 16 of the respondents with high education also had high knowledge, a much greater number than that of the ones having low knowledge, which was simply 2 respondents.

3.4. Bivariate analysis

The bivariate analysis for this inquiry was run by SPSS version 16. To find out the relationship of knowledge and education level to the rationality of self-medication on child diarrhea on the banks of Code River in Jogoyudan Kampong, Yogyakarta, a chi-square test with a confidence interval of 95% was carried out.

3.4.1. Correlation between education level and rationality of self-medication on childhood diarrhea on the Code River Banks in Jogoyudan Kampong, Yogyakarta

Results of data analysis to establish the relationship of education level to the rationality of self-medication on child diarrhea are laid out in Table 13.

Table 13. Correlation between education level and the rationality of self-medication on child diarrhea on the Code River banks in Jogoyudan Kampong, Yogyakarta

Education	Rationality		Chi-Square	OR	CI	P
	Rational	Irrational				
Low	3	19	9.808	0.100	0.021-0.470	0.002
High	11	7				

According to data analysis results in Table 13, the counted chi-square value was 9.808, which was higher than the critical chi-square value with df=1 of 3.841. It also transpired that the p value = 0.002 ($p<0.005$). It could be concluded therefore that there was a relationship between education level and the rationality of self-medication on diarrheic children on the banks of the Code River in Jogoyudan Kampong, Jetis Sub-district, Yogyakarta.

This result conforms earlier investigations [8-10]. One study mentioned that parents' education impacts on their knowledge in taking actions to treat diarrhea. Well-educated parents hold proper knowledge about actions in medicating diarrheic children [9], whilst parents with lesser education lack the same knowledge. Other research also stated that education relates to rationality, in that education is one of the factors in gathering information which influences rationality. The higher the education, the vaster the information acquired, which would be sufficient to avoid irrational medication [10]. The survey undertaken within the proximity of the TPA in Banguntapan Sub-district, Bantul Regency, also yielded that there was a positive correlation between the education level of respondents and the rationality of their self-medication on childhood diarrhea [8].

3.4.2. Correlation between knowledge level and rationality of self-medication on childhood diarrhea on the Code River Banks in Jogoyudan Kampong, Yogyakarta

Data analysis results to identify the relation between knowledge and the rationality of self-medication on child diarrhea are presented in Table 14.

Table 14. Correlation between knowledge level and the rationality of self-medication on child diarrhea on the Code River banks in Jogoyudan Kampong, Yogyakarta

Knowledge	Rationality		Chi-Square	RR	CI	P
	Rational	Irrational				
Low	0	19	19.476	3.00	1.638-5.493	0.000
High	14	7				

Table 14 displays that the calculated chi-square value was 19.476, greater than the critical chi-square value with df=1 of 3.841. It was also revealed that the p value was 0.000 ($p<0.005$). It can thus be determined that respondents with a high knowledge level were able to conduct rational self-medication on childhood diarrhea three times as well as respondents with a low knowledge level.

This finding concurs with three preceding studies [9-11]. Parents' good knowledge relates to proper actions of diarrhea medication, while inadequate knowledge would lead to inappropriate acts of diarrhea medication [9]. Another inquiry cited that knowledge correlates with rationality as a factor that affects rationality. Proper knowledge can avert irrationality in medication in terms of inaccurate dose, interval, and duration [10]. The study carried out near the TPA in Banguntapan Sub-district, Bantul Regency, also divulged a positive relationship between the knowledge level of respondents and the rationality in their self-medication on childhood diarrhea [8].

The limitation of this study was that the responses to the question items in the questionnaires were not followed up by in-depth interviews about self-medication on the respondents' diarrheic children, raising the possibility of factors influencing the questionnaire results that were not identified by the researchers.

4. Conclusions

From 40 respondents, it was found that 14 respondents (35%) performed rational self-medication on childhood diarrhea, whereas 26 respondents (65%) did so irrationally. There was a correlation between education level and the rationality of self-medication on child diarrhea on the banks of Code River in Jogoyudan Kampong, Jetis Sub-district, Yogyakarta, as can be noted from a chi-square value of 9.808 (>3.841) and a p value of 0.002 (<0.05). There was a relationship between knowledge and the rationality of self-medication on childhood diarrhea on the banks of Code River in Jogoyudan Kampong, Jetis Sub-district, Yogyakarta, as denoted by a chi-square value of 19.476 (>3.841) and a p value of 0.000 (<0.05).

Acknowledgements

The authors would like to express their gratitude to the heads of RW 10, 11, 12, and 13 along with the residents around the Code River banks, Jogoyudan Kampong, Jetis Sub-district, Yogyakarta who contributed during the research process until the authors were able to complete this study.

References

[1] Anonymous 2015 *Profil Kesehatan Kota Yogyakarta* Yogyakarta City Health Agency.

[2] Puspitasari D E 2009 Dampak pencemaran air terhadap kesehatan lingkungan dalam perspektif hukum lingkungan (Studi kasus sungai Code di Kelurahan Wirogunan Kecamatan Mergangsan dan Kelurahan Prawirodirjan Kecamatan Gondomanan Yogyakarta) *Mimbar Hukum* **21**(1) 24-33.

[3] Anonymous 2007 *Pedoman Penggunaan Obat Bebas Dan Bebas Terbatas* Ministry of Health RI Jakarta.

[4] Purnasari F D 2013 Hubungan pengetahuan dan tingkat pendidikan terhadap kerasionalan pengobatan pada swamedikasi diare anak di Sekitar Sungai Gajah Wong di Kelurahan Muja Muju Kecamatan Umbulharjo *Bachelor Thesis* Faculty of Pharmacy Universitas Ahmad Dahlan Yogyakarta.

[5] Arikunto S 2013 *Prosedur Penelitian Suatu Pendekatan Praktik* (Jakarta: Rineka Cipta) 211-221.

[6] Notoatmodjo S 2012 *Metodologi Penelitian Kesehatan* (Jakarta: Rineka Cipta) 164, 168.

[7] Trihendradi C 2011 *Langkah Mudah Melakukan Analisis Statistik Menggunakan SPSS 19* (Yogyakarta: Penerbit Andi) 145-147, 215-217.

[8] Noviah 2013 Hubungan pengetahuan dan tingkat pendidikan terhadap kerasionalan penggunaan obat-obatan pada swamedikasi diare anak di sekitar Tempat Pembuangan Akhir (TPA) Kecamatan Banguntapan Kabupaten Bantul *Bachelor Thesis* Faculty of Pharmacy Universitas Ahmad Dahlan Yogyakarta.

[9] Mwambete K D and Joseph R 2010 Knowledge and perception of mother and caregivers on childhood diarrhoea and its management in Temeke Municipality Tanzania *Tanzania J.of Health Res.* **12**(1) 1-9.

[10] Sadikin, ZD 2011 Penggunaan obat yang rasional *J Indon Med Assoc* **61**(4) 146.

[11] Njume, C and Goduka NI 2012 Treatment of diarrhoea in rural african communities: an overview of measures to maximise the medicinal potentials of indigenous plants, *International Journal of Environmental Research and Public Health* 2009 **9**: 3911.

IOP Conf. Series: Materials Science and Engineering **259** (2017) 012016 doi:10.1088/1757-899X/259/1/012016

Formulation and characterization of fisetin nanosuspension

M Dzakwan[1], G E Pramukantoro[1], R Mauludin[2], S Wikarsa[2]

[1]Faculty of Pharmacy, Universitas Setia Budi, Surakarta, Central Java, Indonesia

[2]Department of Pharmaceutics, School of Pharmacy, Institut Teknologi Bandung, West Java, Indonesia

E-mail: mdzakwan9@setiabudi.ac.id

Abstract. Fisetin (3,3,4,7-tetrahydroxyflavone) is a natural antioxidant that has shown to posses anticancer, antioxidant and anti-inflammatory properties. However, the poor solubility leads to poor bioavailability and limits its development. The aim of the study is to investigate the effect of fisetin nanosuspension using a precipitation-sonication method and additional stabilizers polysorbat 80, SLS, F68, PVP, PVA and HPMC on particle size average and the polydispersity index. The suspensions of microcrystalline FIS were prepared by a precipitation method with different proportion of stabilizers fixed. The nanosuspension produced was then characterized using Photon Correlation Spectroscopy (PCS) in term of particle size distribution, polydispersity index and morphology particle (SEM). Result showed fisetin nanosuspension were successfully prepared by anti-solvent precipitation with additional stabilizer SLS and PVA. The nanosuspension containing PVA showed smaller average particle size of 406 nm, a polydispersity index of 0.22±0.1 was obtained. The drug particles precipitated with the PVA as stabilizer were spherical in shape.

Keywords: nanosuspension, fisetin, bottom up, precipitation, ultrasonication

1. Introduction

Flavonoids are a large class of plant secondary metabolites belonging to the wider family of natural polyphenols and generally found in plants, food, fruits and vegetables [1, 2]. They are believed to have beneficial effects against multiple diseases, including cancers, cardiovascular disease, inflammatory, antioxidant and neurodegenerative disorders [3]. Fisetin (FIS, 3,7,3',4'-tetrahydroxyflavone, Figure 1 belongs to the flavonol subgroup of flavonoids found in 65 many vegetables and fruit and is especially rich in apples, strawberries, onions and mangoes [4, 5]. It exhibits antioxidant, antiinflammatory and anticarcinogenic activities. Recently, it was also considered to possess neuroprotective effects against the aging process, cerebral damage and neurodegenerative disorders [5, 6].

IOP Conf. Series: Materials Science and Engineering **259** (2017) 012016 doi:10.1088/1757-899X/259/1/012016

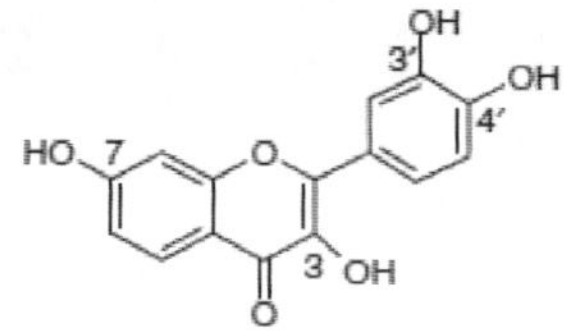

Figure 1. Chemical structure of fisetin [6]

Fisetin has low oral bioavailability (44.1%) probably due to its high lipophilicity (log P 3.2) and low (10.45 μg/mL) aqueous solubility [7]. Moreover, the fisetin molecule bearing 4 hydroxylsubstituents, including a catechol on the phenyl B ring, is extensively metabolised *in vivo* [5]. Consequently, a search for solubilized forms of fisetin is under way, with some promising results obtained by nanoemulsion formation, development of a liposomal formulation, complexation with a cyclosophoroase dimerand nanochelate formation [4-8].

Recently, the nanosuspension technology has been successfully applied to tackle the formulation issue of the poorly soluble drugs. Nanosuspension is a carrier-free colloid drug delivery system containing only minimum stabilizers and pure drug particles with a mean particle size in the nanometer range, typically between 10 and 1000 nm [9, 10]. Nanosuspension properties such as increased drug solubility, high dispersity and homogenization, intravenously injectable, simple production process, universal adaptivity enable its applications in the formulation of large amounts of poorly soluble compounds. Furthermore, the formation of suspensions is one of effective approaches for the stabilization of chemically labile molecules insoluble in aqueous solution [11, 12].

Generally, the nanosuspension techniques are classified as bottom up processes and top down processes according to the differences of the production principle [10]. In the bottom up processes the poorly water-soluble drug is first dissolved in an organic solvent and then precipitated through a non-solvent addition in the present of stabilizers, as in supercritical fluid (SCF) technology, evaporative precipitation into aqueous solution (EPAS), spray-freezing into liquid process, and emulsion-solvent evaporation [13, 14]. These processes were simple and cost effective without any high energy input. However the following prerequisites should be met: (i) the drug should be soluble at least in one solvent. (ii) The solvent is miscible with a non-solvent [4]. The top down methods means the mechanical comminution processes of larger drug particles, as in media milling, microfluidization and high pressure homogenization (HPH) [16]. These methods require no harsh solvents but involve high energy input and low power efficiency [16, 17]. In addition some measures should be taken to minimize the degradation of heat sensitive drugs resulted from the heat generated in the comminution progress [10].

The aim of the present study was to develop fisetin nanosuspension using a precipitation-sonication method and additional stabilizers polysorbat 80, SLS, F68, PVP, PVA, HPMC and Eudragit. The nanosuspension produced was then characterized using photon correlation spectroscopy (PCS) in term of particle size distribution, zeta potential, polydispersity index and morphology particle using scanning electron microscope (SEM).

2. Materials and Methods

2.1. Materials

FIS (98%) raw material was purchased from Shaanxi Dideu Medichem Co. Ltd (Xi'an, China). Polyvinyl alcohol (PVA) were kindly gifted from BASF. Pluronic F68 was purchased from the Sigma-Aldrich Chemical Co. Ltd (USA). Sodium lauryl sulphate (SLS), polyvinyl pyrolidone (PVP) and Hydroxypropylmethylcellulose (HPMC) was obtained from Cognis Indonesia Ltd. Polysorbate 80 (Tween 80®) was obtained from Shino Japan Chemical. All other reagents were of analytical grade.

2.2. Precipitation process (bottom-up method)
Generally, the suspensions of microcrystalline FIS were prepared by a precipitation method with different proportion of stabilizers fixed. The suspensions were prepared by dissolving 0.01 g of FIS in 2.0 mL of ethanol as the organic phase. Then the organic phase was injected slowly at a stirring rate of 600 rpm using a magnetic stirrer (RCT Basic, IKA, Staufen, Germany) into 20 mL of aqueous phase containing polysorbat 80, SLS, PVA, F68, PVP and HPMC (Table 1). During this process, the temperature was controlled using an ice water bath. Low temperature (below 8°C) was maintained through out the process using an ice water bath which controlled the precipitation rate. The treatment was done with an ultrasonic probe sonicator (Q Sonica Sonicator, USA) at ultrasonic power input of 50% for 8 min length. The probe with a tip diameter of 8 mm was immersed 10 mm in the liquid resulting in the wave traveling downwards and reflecting upwards. The suspensions were kept under vacuum at room temperature for 1 hour to remove ethanol.
2.3. Particle size analyses
The particle size and zeta potential value of FIS nanosuspensions were determined by the DelsaTM Nano C Particle Analyzer (Beckman Coulter GmbH, Germany). The samples were adequately diluted with deionised water and placed in an electrophoretic cell. The measured parameters by PCS are the average particle size diameter (ZAve) and the polydispersity index (PI). The mean particle size of three measurements was taken.
2.4. Scanning Electron Microscopy (SEM)
The morphologies of raw FIS and nano-sized FIS were examined using a scanning electron microscope (JEOL JSM-7001F, Japan) operated at an accelerating voltage of 15 kV and a secondary detector. Freshly prepared FIS nanosuspensions and dispersions of raw FIS were deposited on a glass slides following the evaporation of solvent.

Table 1. Composition of FIS nanosuspension

	Formulation (%)				
	A	B	C	D	E
FIS	1.0	1.0	1.0	1.0	1.0
PVA	1.5	-	-	-	-
SLS	-	0.5	-	-	-
Tween 80	-	-	0.5	-	-
HPMC	-	-	-	1.0	-
Eudragit	-	-	-	-	1.0
Water	97.5	98.5	98.5	98.0	98.0

3. Results and Discussion
3.1. Preparation of nanosuspensions
FIS nanosuspensions were produced by precipitation–ultrasonication technique. The aqueous phase containing suitable stabilizer has been used as antisolvent and the use ethanol and acetone as water miscible solvent that having good solubility FIS. Acetone and ethanol have been tried to serve as solvent for the preparations. But acetone-based formulations showed aggregation of particles and settling on storage. The viscosity of the preparation increased indicated flocculation behaviour. Hence acetone has not been used for further studies. Ethanol as solvent produces very uniform-sized nanoparticles on precipitation. The effect of various variables like type of stabilizer, concentration of stabilizer, stirring time and ultrasonication has been observed.

From the preliminary studies no specific effect on particle sizes has been observed at different effect of stirring time (5, 10, 30, 60 min) on the particle size studied. No sign of aggregation due to stirring has been observed, and the particle size did not depend on stirring time. SLS, polysorbat 80, PVA, HPMC and eudragit, at various concentrations (0.5, 1.0 and 1.0%), had been used as stabilizer for the preparations. From the preliminary studies, 0.5% and 1.5% concentration was found to be optimum for both SLS and PVA, respectively (Figure 2). The PVA-based formulations at concentration 1.5% gave for smaller and more uniform nanoparticles and more stable at storage under room temperature as long four weeks than SLS as stabilizer were a bit bigger.

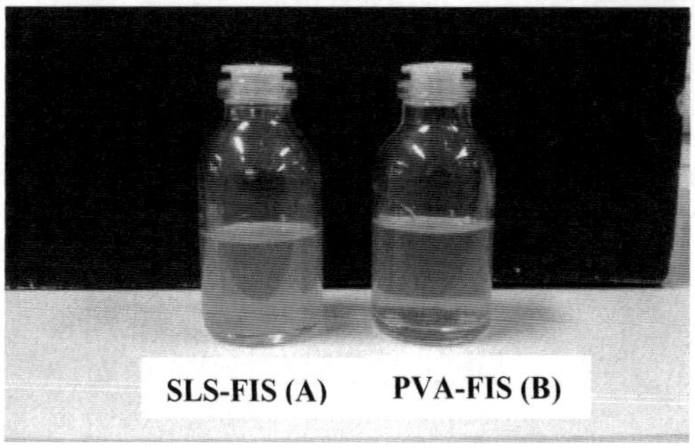

Figure 2. Appearance of fisetin nanosuspensions (A) SLS-Fisetin and (B) PVA-Fisetin

The effect of ultrasonication has been studied since the preparation obtained by precipitation alone showed less homogeneity and particles were obtained in range with average particles size at 406–997 nm. The particle size was considerably decreased and more uniform in case of precipitation followed by ultrasonication for 8 min under cold condition. Hence, precipitations with ultrasonication have been used for further preparations of nanoparticles of FIS.

3.2. Size measurement and polydispersity index

The particle size and the polydispersity Index (PI) of the formed drug particles was measured immediately after precipitation and sonication by dynamic laser light scattering (DelsaTM Nano C Particle Analyzer Beckman Coulter GmbH, Germany). The average particle size of FIS nanoparticles was found to be in the range of 406–997 nm (Table 2). The formulations containing 1.5% PVA as stabilizer showed particle size in the range 406 nm smaller than others stabilizer. The formulations were homogeneous as indicated by polydispersity index of 0.220±0.1. The particle size distributions were found to be more uniform as the polydispersity index narrowed down to 0.3–0.5.

Table 2. Input power, mean of PCS particle size and PI FIS nanosuspension

Formulation Code	Parameters		
	Input power (%)	Z-average (nm)	PI
A	50	406	0.220
B	50	574	0.291
C	50	767	0.236
D	50	849	0.201
E	50	997	0.308

3.3. Scanning Electron Microscopy (SEM)

Morphology of precipitated drug particles in the suspension after air drying followed by oven-drying. The drug particles precipitated with the PVA as stabilizer were spherical in shape and the size ranges 406 nm was shown in Figure 3. The particles were discrete and uniform in size and there was no sign of agglomerations. The drug particles precipitated with the SLS as stabilizer were slightly bigger and spherical and somewhat cuboidal and the size ranges 575 nm.

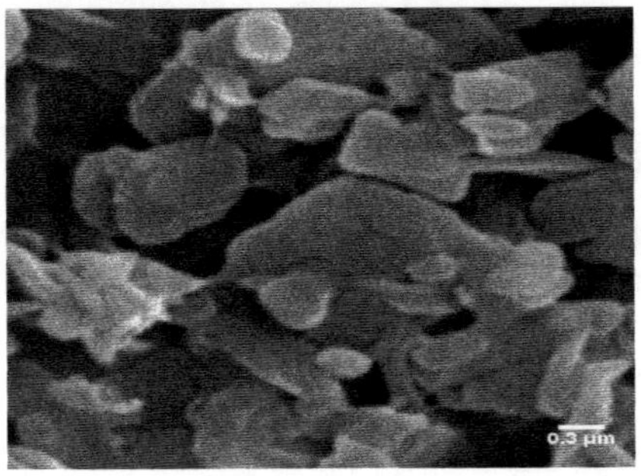

Figure 3. SEM photomicrograph of PVA based Fisetin (X 10.000)

4. Conclusions

It may be concluded that nanoprecipitation with ultrasonication had potential to formulate homogenous nanosuspensions with uniform-sized stable nanoparticles of fisetin. The preparations were found to be physically stable with PVA and SLS as stabilizer.

Acknowledgement

The authors would like to thank to Ministry of Research, Technology and Higher Education of Indonesia who grant this research through PEKERTI research scheme.

References

[1] Leonarduzzi G, Testa G, Sottero B, Gamba P and Poli G 2010 *Current Medicinal Chemistry.* **17** 74-95.
[2] Chen P Y, Ho Y R, Wu M J, Huang S P, Chen P K, Tai M H, Ho C T and Yen J H 2014 *Food& Function* 1-10.
[3] Burda S and Oleszek W 2001 *J. Agric. Food Chem.* **49** 2774-9.
[4] Sowa M, Slepokura K and Jon E M 2014 *Cryst. Eng. Comm.* **16**(46) 10592-10601.
[5] Wang L, Zhang D Z and Wang Y X 2016 *Pham. Res.* **2** 5-9.
[6] Ragelle H, Manciet S C, Seguin J, Brossard D, Scherman D, Arnaud P, Guy G and Chabot G G 2012 *Int. J. Pharm.* 452–9.
[7] Bothiraja C, Yojana, B D, Pawar A P, Shaikh K S and Thorat U H 2014 *Expert Opin. Drug Deliv* **11**(1) 17-29.
[8] Guzzo M R, Uemi M, Donate P M, Nikolaou S, Antonio Eduardo H, Machado A E H and Okano L T 2006 *J. Phys. Chem. A.* **110**(36) 10545-10551.
[9] Gao L, Zhang D R and Chen M H 2008 *J. Nanopart. Res.* **10** 845–862.
[10] Keck C M and Muller R H 2006 *Eur. J. Pharm. Biopharm.* **62**(1) 3-16.
[11] Schwitzera J M, Achleitner G, Pomperb H and Müller R H 2004 *Eur. Pharm. Biopharm.* **58** 615–619.
[12] Pu X, Sun, J, Wang Y, Liu X, Zhang P, Tang X, Pan W, Han J and He Z 2009 *Int. J. Pharm.* **379** 167–173.
[13] Li X S, Wang J X , Shen Z G, Zhang P Y, Chen J F and Yun J 2007. *Int. J. Pharm* **342** 26–32.
[14] Rabinow B E 2004 *Nature Rev. Drug. Disc.* **3** 785–796.
[15] Van Eerdenbrugh B, Van den Mooter G and Augustijns P 2008 *Int. J. Pharm.* **364**(1) 64-75.
[16] Patravale V B, Date A A and Kulkarni R M 2004 *J. Pharm. Pharmacol* **56** 827–840.
[17] Liversidge E M, Liversidge G G and Cooper E R 2003 *Eur. J. Pharm. Sci.* **18** 113–120.

Total alkaloid content in various fractions of *Tabernaemonata sphaerocarpa* Bl. (Jembirit) leaves

N Salamah, D S Ningsih

Faculty of Pharmacy, Universitas Ahmad Dahlan, Yogyakarta, Indonesia

E-mail: ninasalamah1996@gmail.com

Abstract. *Tabernaemontana sphaerocarpa* Bl. (Jembirit) is one of the *Apocynaceae* family plants containing alkaloid compound. Traditionally, it is used as an anti-inflammatory medicine. It is found to have a new bisindole alkaloid compound that shows a potent cytotoxic activity in human cancer. This study aimed to know the total alkaloid content in some fractions of ethanolic extract of *T. sphaerocarpa* Bl. leaf powder was extracted by maceration method in 70% ethanol solvent. Then, the extract was fractionated in a separatory funnel using water, ethyl acetate, and hexane. The total alkaloid content in each fraction was analyzed with visible spectrophotometric methods based on the reaction with Bromocresol Green (BCG). The total alkaloids in water fraction and ethyl acetate fraction were (0.0312±0.0009)% and (0.0281±0.0014)%, respectively. Meanwhile, the total alkaloid content in hexane was not detected. The statistical analysis, performed in SPSS, resulted in a significant difference between the total alkaloids in water fraction and ethyl acetate fraction. The total alkaloid in water fraction of *T. sphaerocarpa* Bl. was higher than the one in ethyl acetate fraction.

Keywords: *T. sphaerocarpa* Bl., alkaloid, visible spectrophotometry, fractionation

1. Introduction

Alkaloid is a secondary metabolite compound that has been providing more benefits to the world of medicine and pharmaceutical dosage form compared to the other metabolites. As a type of secondary metabolites, it shows a wide biological activity. It occurs in plants, fungi, bacteria, amphibians, insects, marine animals, and humans. Plants and fungi, which are rich in natural materials, were used as pain relief in ancient times [1]. The *Taebernaemontana* species from the *Apocynaceae* family particularly has copious indole alkaloids. This type of alkaloid is an indicator compound that offers a wide range of benefits and plays an immense role in certain species classification [2].

The research on *T. sphaerocarpa* Bl stem [3] found two new bisindole alkaloid compounds, namely biscarpamontamine A, having an aspidosperma-iboga-type skeleton, and biscarpamontamine B, possessing an aspidosperma-aspidosmema-type skeleton. Biscarpamontamine A shows potent cytotoxicity against various types of human cancer cells.

The isolation of alkaloids from plant sources needs a good solvent to extract them, i.e., organic solvents such as ether, alcohol, and benzene. Fractionation is a further isolation of chemical constituents from the crude extracts to obtain purer compounds and hence, accentuate their influence on a specified bioassay test [4]. The major chemical property of alkaloids is their basic nature, which causes them to easily decompose following exposure to heat and light in the presence of oxygen. Physically, alkaloids are colorless. The free alkaloidal base is soluble only in an organic solvent, whereas alkaloidal salts are highly soluble in water [5].

A preliminary detection of alkaloids is achievable using Mayer's and Dragendorff's reagents. Mayer's reagent contains potassium iodide and mercuric chloride, while Dragendorff's reagent consists of bismuth nitrate and potassium iodide in a solution of nitric acid [6]. Alkaloids react with Mayer's reagent and produce cream-colored precipitate. Furthermore, Dragendorff's reagent detects a positive presence of alkaloid from the formation of brown to orange precipitate and the absence of such precipitate when added with H_2SO_4 [7].

The determination of total alkaloids using a visible spectrophotometric method with Bromocresol Green (BCG) is a simple and sensitive technique that requires no special equipment. BCG can react with certain alkaloids, i.e., the ones that have nitrogen inside their structure, but not with amine and amide alkaloids. The reaction of alkaloids with BCG forms a yellow-colored product [8].

Based on the previous research on the effects of extraction methods on the number of total alkaloids in *T. sphaerocarpa* Bl leaf, maceration produces higher total alkaloids and is thereby a better extraction technique than Soxhlet extraction [7]. Fractionation is a method used to separate one component from the others or to break a mixture into its constituent parts. The fractionation of a component can be carried out using chromatography. The isolation of alkaloids from plant sources needs a good solvent to extract them, i.e., organic solvents such as ether, alcohol, and benzene [4]. Alkaloids are separated from crude drugs using an extraction technique with non-aqueous organic solvent. Most alkaloids are present in plants in the form of salts soluble in methanol or ethanol [9].

2. Materials and Methods

2.1. The determination of T. sphaerocarpa Bl. leaf

The determination took place in the Biology Laboratory, Faculty of Mathematics and Natural Sciences, Universitas Ahmad Dahlan, Yogyakarta using *T. sphaerocarpa* Bl. leaf.

2.2. Sample processing

The collected *T. sphaerocarpa* Bl. leaves were washed, and then dried in an oven at a temperature of 60^0C. After drying, the leaves were powdered using a grinder and passed through a multi-stage sieve to obtain 50/70 mesh particle size.

2.3. Extract preparation

The extract was prepared using maceration with three replications. Each extraction cycle used 150 gram of 50/70-mesh powder of *T. sphaerocarpa* Bl. leaf that was macerated with 500 mL of 70 % v/v ethanol. This 24-hour maceration required a 3-hour stirring first. It was repeated three times until the alkaloids were extracted perfectly. The decision of a perfect extraction was based on a qualitative observation on the macerate. Furthermore, the macerate was filtered with Buechner funnel under vacuum. The remaining solvents in it were removed using a rotary evaporator to obtain a viscous extract, which was tested qualitatively using Mayer's and Dragendorff's reagents. A positive reaction of alkaloids with Dragendorff's reagent is shown by the presence of orange-colored precipitate. Meanwhile, the presence of alkaloids is evidenced by the resultant cream-colored precipitate.

2.4. Non-specific standardization

2.4.1. Chemical yield calculation

The chemical yield was obtained by measuring the final weight (viscous extract) and comparing it with the initial weight of the powder before extraction.

2.4.2. Loss on powder drying and extract's water content

The loss on drying and the moisture content of the extract were determined with Halogen Moisture Analyzer. This instrument was positioned horizontally by adjusting the leveling screw. It is perfectly horizontal when the air bubble in the water pass is in the center. Then, the Moisture Analyzer Balance was switched on and left for 30 minutes before using. The red button was pressed until the display was on. After the start button on the display was pressed, the heating module was opened to load the pan. It was then closed, and the zero button was pressed to set the Moisture Analyzer to zero. Afterward, the heating module was opened again, a sample of at least 0.5 grams of *T. sphaerocarpa* Bl. leaf extract was placed on the pan, and the heating module of the moisture analyzer was closed.

2.5. Qualitative analysis of alkaloids

The extract was dissolved in 3.0 mL of 70% v/v ethanol and added with 5.0 mL of HCl 2 M and 0.5 g of NaCl. This filtrate was then filtered, added with 3 drops of HCl 2 M, and divided into 4 tubes. Filtrate A functioned as a blank solution. Meanwhile, filtrate B was added with Mayer's reagent, filtrate C was added with dilute H_2SO_4, and filtrate D was added with Dragendorff's reagent [10].

2.6. Fractionation and the determination of total alkaloids

Five grams of the viscous extract were dissolved in 10 mL of water. Sonication was performed and a little of ethanol was added only when necessary (to help the dissolution process). This solution was then poured into a separating funnel, added with 10 mL of hexane, and shaken to produce a solution with two phases, namely water and hexane. These two phases were separated and collected. The extraction with hexane solvent was carried out until the hexane phase had the same color as the hexane solvent. The water phase was collected for fractionation using 10 mL of ethyl acetate. The fractionation was repeated until the ethyl acetate phase has the same color as the original ethyl acetate solution. The total alkaloids in every fraction, namely water, ethyl acetate, and hexane, were analyzed with UV-Vis spectrophotometric method.

2.6.1. The determination of operating time

A strychnine solution with a concentration of 25 μg/mL was poured into a separating funnel, added with 5.0 mL of phosphate buffer solution at pH 4.7 and 5.0 mL of BCG solution, and extracted with 5.0 mL of chloroform (two times). The chloroform phase was then taken. The extraction results were collected in a 10.0 mL volumetric flask and added with chloroform up to the mark. Then, the stable absorption time was checked at a wavelength of 470 nm. The data obtained from this analysis was used to make a curve representing the relationship between absorbance and time.

2.6.2. The determination of wavelength at maximum absorbance

A strychnine solution with a concentration of 50 μg/mL was poured into a separating funnel, added with 5.0 mL of phosphate buffer solution at pH 4.7 and 5.0 mL of BCG solution, and extracted with 5.0 mL of chloroform (two times). The chloroform phase was then taken. The extraction results were collected in a 10.0 mL volumetric flask and added with chloroform up to the mark. The absorbance of this solution was then checked at a wavelength of 350-700 nm using the operating time obtained from the above analysis. The resultant data in this analysis was used to make a curve representing the relationship between absorbance and wavelength.

2.6.3. The determination of standard curve

A standard strychnine solution with a concentration of 50 μg/mL was pipetted out by 1.0, 2.0, 3.0, 4.0, 5.0, 6.0, and 7.0 mL into a separating funnel, added with 5.0 mL of phosphate buffer at pH 4.7 and 5.0 mL of BCG solution, and extracted with 5.0 mL of chloroform (two times). The chloroform phase was then taken. The extraction results were collected in a 10.0 mL volumetric flask and added with chloroform up to the mark. Afterward, their absorbance was checked against the maximum wavelength.

2.6.4. Sample preparation

The water and ethyl acetate fractions, whose solvents had been evaporated, were added with 2N HCl until dissolved perfectly. Then, 1.0 mL of filtrate was extracted using 5.0 mL of chloroform (three times). Afterward, the pH of the solution was neutralized using 0.1N NaOH and added with 5.0 mL of phosphate buffer solution at pH 4.7 and 5.0 mL of BCG solution. This mixture was then extracted again using 5.0 mL of chloroform (two times). The chloroform phase was collected in a 10.0 mL volumetric flask and added with chloroform up to the mark. The absorbance at the maximum wavelength was considered in the calculation of the total alkaloid content.

3. Results and Discussion

The determination of stem, leaf, and stalk took place in the Biology Laboratory, Faculty of Mathematics and Natural Sciences, Universitas Ahmad Dahlan, Yogyakarta. The results proved that the plant used in this research was jembirit. It is *T. sphaerocarpa* Bl. species from the family *Apocynaceae* and the genus *Tabernaemontana*.

The water content of the powder of *T. sphaerocarpa* Bl. leaf in the three replications was 4.72%, 4.31%, and 8.32%, respectively. The data is presented in details in Table 1.

Table 1. The weight loss of *T. sphaerocarpa* Bl. leaf powder in drying

Replications	Weight of powder (gr)	Loss on Drying (%)	Mean±LE (%)	CV (%)
1	0.508	4.72	4.56±0.544	4.809
2	0.511	4.31		
3	0.510	4.65		

It shows that the weight loss of the powder of *T. sphaerocarpa* Bl. leaf in drying meets the requisites of Indonesian Herbal Pharmacopoeia (IHP) i.e., less than 10%. This percentage ensures that the powder can be stored and preserved from molds and other fungi. The chemical yield of the ethanolic extract of *T. sphaerocarpa* Bl. leaf is presented in Table 2.

Table 2. The chemical yield of the ethanolic extract of *T. sphaerocarpa* Bl. leaf

Replications	Crude drug (g)	Weight of extract (g)	Chemical Yield (%)	Mean±LE (%)
1	150.09	34.8	23.186	22.364±2.154%
2	150.12	33.7	22.448	
3	150.06	32.2	21.458	

Chemical yields signify the effectiveness of an extraction process in an analysis. The effectiveness is influenced by various factors, namely extraction solvent, stirring speed, the particle size of the powder, and the length of the extraction process. The calculation results showed that the mean of the chemical yield was 22.364%.

The moisture content was measured by the amount of water found in the material. In this research, it was measured to provide the maximum limit or range of the water content in the material [11]. The water content of the extract is summarized in Table 3.

Table 3. The determination of the water content in the ethanolic extract of *T. sphaerocarpa* Bl. leaf

Replications	Weight of powder (g)	Loss on Drying (%)	Mean±LE (%)	CV (%)
1	0.508	10.63	10.203±0.956%	3.773
2	0.506	10.10		
3	0.516	9.88		

Materia Medika Indonesia (MMI) explains that the water content of a sample is never less than 10% unless stated otherwise. A viscous extract containing 30% water is susceptible to the growth of microorganisms [12]. The water content in a pharmaceutical dosage form largely depends on extraction solvent, evaporation time, and storage conditions. In this research, the water content exceeded the upper limit of the allowed presence set by MMI possibly because of the solvent used in the extraction and the loosely sealed container that stored the extract.

The qualitative analysis of the ethanolic extract of *T. sphaerocarpa* Bl. leaf resulted in the presence of alkaloids. The positive result was evidenced by the cream-colored precipitate, i.e., from the reaction of alkaloids with Mayer's reagent, and the brick-red one, i.e., the response of alkaloids to Dragendorff's reagent. The whitish precipitate from the usage of Mayer's reagent is a potassium-alkaloid complex.

The determination of total alkaloids using a visible spectrophotometric method with Bromocresol Green (BCG) is a simple and sensitive technique that requires no special equipment. BCG can react with certain alkaloids, i.e., the ones that have nitrogen inside their structure, but not with amine and amide alkaloids. The reaction of alkaloids with BCG forms a yellow-colored product, as presented in figure 1[8].

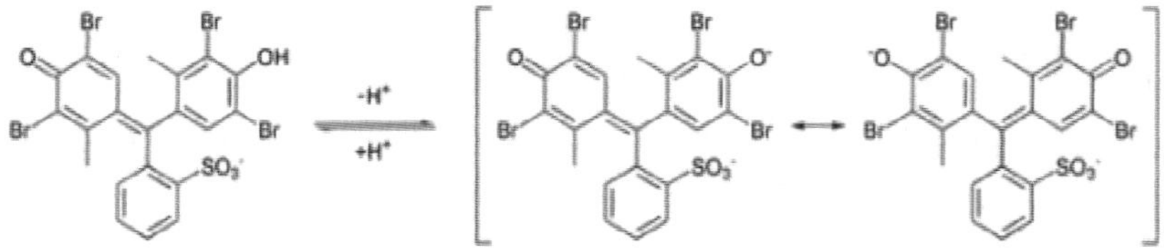

Figure 1. The structure and resonance of bromocresol green

The determination of the operating time (OT) of the samples was performed because the exact alkaloid content in *T. sphaerocarpa* Bl. was unknown. The results showed that the standard OT was 34-37 minutes, while the OT of the samples was 37-39 minutes. Furthermore, they also revealed that the maximum wavelength of the BCG complex with strychnine was 416 nm, while that of the BCG complex with alkaloids in *T. sphaerocarpa* Bl. leaf was 412.8 nm. The curve for the standard strychnine was measured at a maximum wavelength of 416 nm and within an operating time of 43-47 minutes. The absorbance obtained from the series of the standard strychnine solution is presented in Table 4.

Table 4. The absorbance of standard strychnine in different concentrations

Concentrations (μg/mL)	Absorbance
5	0.276
10	0.348
15	0.446
20	0.502
25	0.638
30	0.662
35	0.770

The equation obtained from the standard curve is y=0.1644x+0.1914 where x is content (μg/mL), and y is absorbance. The r-value, calculated from the standard curve, is 0.9941. The quantitative analysis using spectrophotometry provided a dataset of the absorbance of each sample, as presented in Table 5. The absorbance levels of the hexane fraction samples were 0.005, 0.062, and 0.016. Because these samples had a low absorbance, the quantification of the alkaloid content in these samples was not performed. The calculation results, as seen in Table 5, showed that the total alkaloid content in water fraction was higher than the one in ethyl acetate fraction.

Table 5. The calculation results of the determination of total alkaloid content in *T. sphaerocarpa* Bl. leaf

Fractions	Replications	Sample weights (g)	Absorbance	Total Alkaloid Content (%)	X ± LE (%)	CV (%)
Water	1	4.9984	0.700	0.0310	0.0312±0.0009	1.2820
	2	5.0037	0.711	0.0317		
	3	5.0141	0.702	0.0310		
Ethyl acetate	1	4.9984	0.310	0.0289	0.0281±0.0014	2.1352
	2	5.0037	0.306	0.0279		
	3	5.041	0.305	0.0276		

4. Conclusions

The total alkaloid content in water fraction was (0.0312±0.0009)%, while the total alkaloid in ethyl acetate fraction was (0.0281±0.0014)%. However, the absorbance of alkaloids in hexane fraction was not detected by the visible spectrophotometric method. The fractionation using water solvent contained higher total alkaloids than using ethyl acetate.

References

[1] Michael H, Joanne B, Simon G, and Elizabeth M W 2010 *Farmakognosi dan Fitoterapi* translated by Syarief W R, Cucu A, Ella E, Euis R F (Jakarta: EGC) 104-105.

[2] Cecilia S M, Jucimar J S, Leda M, Raimundo B F, and Ivo J C V 2005 A new indole alkaloid isolated from *Tabernaemontana hystrix* Steud (Apocynaceae). *Setor de Química de Produtos Naturais, Universidade Estadual Brazil* **16** (6B) 1331-1335.

[3] Zaima K, Hirata T, Hosoya T, Hirasawa Y, Koyama K, Rahman A, Kusumawati I, Zaini N C, Shiro M, and Morita H 2009 *Biscarpamontamines A and B, and Aspidosperma-iboga Bisindole Alkaloid and an Aspidosperma-aspidosperma Bisindole Alkaloid from Tabernaemontana sphaerocarpa* Tokyo Japan Faculty of Pharmaceutical Sciences Hoshi University Ebara 2-4-41.

[4] Ayuni N P S and I Nyoman S 2013 *Isolasi dan Identifikasi Senyawa Alkaloid pada Biji Mahoni (Swietenia mahagoni Jacq.)* Fakultas MIPA Universitas Pendidikan Ganesha Singaraja.

[5] Robinson T 1995 *Kandungan Organik Tumbuhan Tinggi* Edisi VI, translated by Padmawinata K (Bandung: Institut Teknologi Bandung Press) 284-285.

[6] Cordell G A 1981 *Introduction to Alkaloids a Biogenetic Approac*, University of Illinois, Chicago 26-27.

[7] Salamah N, Rozak M, and Al Abror M 2014 Pengaruh metode penyarian terhadap kadar alkaloid total daun Jembirit (*Tabernaemontana sphaerocarpa* Bl.) dengan metode spektrofotometri visibel, *Pharmaciana,* **7(1)** 113-122.

[8] Shamsa F, Hamidreza M, Rouhollah G and Mohammadreza V 2008 *Spectrophotometric Determination of Total Alkaloid in Some Iranian Medicinal Plant* Tehran University of Medical Sciences Tehran Iran **32** 17-20.

[9] Mursyidi A and Sardjoko 1990 *Analisis Metabolit Sekunder* Pusat Antar Universitas Bioteknologi Universitas Gadjah Mada Yogyakarta 71-86.

[10] Dayanti R and Suyatno 2012 Aktivitas antioksidan ekstrak metanol bagian batang tumbuhan paku *Nephrolepis radicans* (Burm.) Kuhn, *UNESA Journal of Chemistry.* **1**(1) *86-92*

[11] Departemen Kesehatan RI 2000 *Parameter Standar Umum Ekstrak Tumbuhan Obat* Direktorat Jenderal Pengawasan Obat dan Makanan Direktorat Pengawasan Obat Tradisional Jakarta 9-11, 10.

[12] Voigt R 1995 *Buku Pelajaran Teknologi Farmasi* translated by Soendani NS and Mathilda BW (Yogyakarta: Gadjah Mada University Press) 561.

The efficacy of black cumin seed (*Nigella sativa*) oil and hypoglycemic drug combination to reduce HbA1c level in patients with metabolic syndrome risk

P N R Rachman[1], Akrom[1,2], E Darmawan[1]

[1]Faculty of Pharmacy, Universitas Ahmad Dahlan, Yogyakarta, Indonesia

[2]Center for Drug Information and Study, Universitas Ahmad Dahlan, Yogyakarta, Indonesia

E-mail: akmaa_uad@yahoo.co.id

Abstract. Metabolic syndrome is a conditions caused by metabolic abnormalities include central obesity, atherogenic dyslipidemia, hypertension, and insulin resistance. HbA1c examination is required to study the long-term glycemic status and to prevent diabetic complications of metabolic syndrome. The purpose of this study is to determine the efficacy of black cumin seed (*Nigella sativa*) oil and hypoglycemic drug combination to reduce HbA1c level in patients with metabolic syndrome risk. This research performed using an experimental randomized single - blind controlled trial design. A total of 99 outpatients at the Jetis I Public Health Center, Yogyakarta, Indonesia with metabolic syndrome risk were divided into three groups: The control group received placebo and two treatment groups received black seed oil orally at dose of 1.5 mL/day and 3 mL/day, respectively, for 20 days. The clinical conditions such as blood pressure, pulse rate, BMI, blood glucose serum and HbA1c levels were examined on day 0 and 21. The results obtained were analyzed with one-way ANOVA test. The mean of HbA1c levels of all groups before treatment was higher than the normal values and there was no significant difference in HbA1c value on day 0. Administration of 1.5 and 3 mL/day of black seed oil for 20 days decreased ($p<0.05$) HbA1c levels. It can be concluded that administration of black cumin seed oil and hypoglycemic drug combination for 20 days in patients at risk of metabolic syndrome may reduce to HbA1c levels.

Keywords: metabolic syndrome outpatients, black cumin seed oil, HbA1c, clinical test

1. Introduction

One of the public health problems in Indonesia is metabolic syndrome (MS). MS is a major risk factor for vascular and degenerative diseases, including cancer [1-3]. In Indonesia, MS prevalence varies based on age, gender, and region between 11 and 24.7% [4, 5]. One of the MS diseases, diabetes mellitus, is associated with the complications of various chronic and fatal diseases. Diabetes complications cause 50% and 30% of deaths due to coronary heart disease and kidney failure respectively. Diabetes also causes disability, in which 30% of patients experienced blindness due to retinopathy complications and 10% of them had to undergo leg amputation [6]. Diabetes mellitus is also associated with sudden death incidence in the community [7]. Epidemiologically and clinically, it has been demonstrated that angiopathic complication is associated with high hemoglobin A1c (HbA1c) levels.

HbA1c levels considered safe are about 6.5-7%. It is one of the glycosylated proteins to form HbA1c. Glycosylation occurs spontaneously in the circulation and will increase when blood glucose levels are high [2]. HbA1c was the average blood glucose level for 2-3 months [2, 8]. More than 50% of diabetes mellitus patients experience failure to control blood glucose levels that lead to HbA1c levels abnormalities (>7%) [9]. High blood glucose or HbA1c levels causes oxidative stress that damages blood vessels and triggers vascular endothelial death [10, 11]. HbA1c is formed from a bond of glucose with an amide group on the valine amino acid at the end of the Hb globulin beta chain. If this condition is prolonged, *irreversible advanced glycosylation end products* (AGE) will be formed on the walls of blood vessels that may be potentially pathogenic and lead to diabetic complications [12, 13]. Increased AGE results in cellular and tissue oxidative stress, causing damage to tissues and blood vessel cells, especially the endothelium tissue and generating an inflammatory reaction [10, 14].

Oral hypoglycemic drugs (OHOs), which are standard therapy for controlling blood glucose in DM patients, are estimated to be only capable of lowering HbA1c by approximately 0.5-2%, depending on the type of individual OHO drugs [9]. Therefore, it is often necessary for additional or complementary therapies to improve the achievement of ideal HbA1c levels of <7% [8, 15]. It is expected that a combination of high antioxidant food diet and OHOs therapy will lower HbA1c levels [12, 16]. Black cumin seed (*Nigella sativa*) oil (BCSO) has empirically been used as a dietary supplement, antioxidant medicinal ingredients, and immunomodulators. BCSO contains thymoquinone, nigellone, thymohydroquinone, dithymoquinone, thymol, carvacrol, nigellcine, nigellimine-N-oxide, nigellidine and alpha hedrin which are empirically and laboratorically antioxidative [17]. In preclinical studies, it has been demonstrated that BCSO administration on Danforth's short tail (SD) mice induced by alloxan can prevent damage to the pancreas and the liver by increasing the p53 gene expression [18]. It is safe and well-tolerated in phase 1 trial [19]. BCSO administration as an adjunctive therapy in hyperlipid and hyperglycemic patients at hospital could significantly reduce total cholesterol, LDL, and fasting blood glucose levels. However, the effect of BCSO on HbA1c in patients with risk of metabolic syndrome at Public Health Center has never been investigated. Therefore, this study aims to investigate the effect of black cumin seed oil on HbA1c in Public Health Center outpatients with the risk of metabolic syndrome [20].

2. Materials and Methods

This study was a randomized control trial with two dose levels. The number of samples was determined according to the phase 2 clinical trial guidelines, where the sample number was 80-300 volunteers. A total of 99 subjects with metabolic syndrome were divided into three groups, 33 subjects of each which would receive BCSO at dose of 1.5 mL/day and 3 mL/day for 20 days.

Research subjects were outpatients at risk of metabolic syndrome at Jetis I Public Health Center, Bantul, Yogyakarta, Indonesia. The inclusion criteria were male or female patients aged >18 years old, willing to become research volunteers proven by voluntarily signing of informed consent, having been diagnosed with diabetes mellitus or dyslipidemia or hypertension with or without central obesity and having become the outpatient at Jetis 1 Public Health Center at least for three months. Exclusion criteria were drop out volunteers, pregnant women, taking NSAID medications and immunomodulatory supplements, allergic with black cumin seed oil, having a history of chronic kidney disease and cancer, undergoing hyperthyroid and diabetic ulcers therapy and active pulmonary TB patients. For 20 days, groups 1 and 2 were given 1.5 mL/day and 3 mL/day of BCSO soft capsule, respectively, while group 3 acts as a placebo group. This clinical trial research protocol has been reviewed and endorsed by the medical research ethics committee of the Faculty of Medicine and Health Sciences, Universitas Muhammadiyah Yogyakarta, Indonesia. The selection of prospective research subjects was based on outpatient computer database at the Public Health Center. From the database, it was found that from May to October 2016, there were 650 patients at risk of metabolic syndrome. The patients' medical records containing information about hypertension, dyslipidemia, or diabetes mellitus with or without obesity were listed and screened to ensure that the medical records data met the inclusion criteria. A total of 120 patients met the criteria, and they were subsequently designated as subject candidates. All prospective subjects were gathered and briefed on general research objectives and benefits and the specific benefits and consequences of participating as research volunteers. Candidates who were willing to participate in the study were required to give consent by signing the informed consent. A total of 112 prospective subjects who have signed then informed consent then underwent physical and laboratory checks. Based on the baseline examination, 13 patients met the exclusion criteria.

The research was conducted at Jetis I Public Health Center, Bantul, Yogyakarta for 10 months from March to December 2016. The research instrument used was a spectrophotometer to measure blood glucose level, while the test material used was BCSO soft capsules containing standardized black cumin seed oil with a composition of: 2.72% thymoquinone and 0.15% caprylic fatty acid, caprat 0.1%, laurat 0.18%, miristat 12.27%, palmitate 0.28%, stearate 7.99%, oleate 0.07%, linoleat 2.85%, eicosanoate 3.15%, eicosenat 0.25%, eicosadienoate 0.03%, arachidonate 0.03%, eicopentanoate 0.03%, behenat 0.06%, docohexanoate 0.04%, 0.02% teracosanoate and placebo in the form of empty capsules.

A total of 99 subjects were divided into three simple randomized groups (with a lottery system). Randomization was done manually by comparison between groups of 1:1:1. Random grouping was done by pharmacist with multiple lot sweepstakes. All subjects were recorded based on sequence number and then divided into three groups: treatment group 1, treatment 2 and placebo group. The drawing was done by taking number 1, 2, and 3. The number obtained by the first drawer acts as the treatment group 1, the number obtained by the second drawer as the treatment group 2, and the number obtained by the last drawer as the treatment group 3. The members of each group were selected using a multiple of 3 of the lottery number obtained. If the first lottery gets the number 1, then the members of the treatment group 1 were 4, 7, 10 and so on until the quota of 33 patients was reached. The same method applied to treatment group 2 and treatment group 3 (placebo).

The treatment given to the subjects was soft capsule (SC) of BCSO with doses of 1.5 and 3 mL/day or placebo. BCSO capsules were prepared by the pharmaceutical industry with *Good Manufacturing Practices* (GMP) certificate. The material was tested to examine the heavy metal content and standardization of the thymoquinone content. The capsule administration was done by doctors at Jetis I Public Health Center. The examination of post-treatment clinical conditions was done also by the doctors every two weeks at Public Health Center. Specifically trained teams were required to make visits to patients' homes twice per week to check blood pressure, pulse, and body mass index, and monitor unexpected reactions and medication compliance. The analysts and research personnel have been trained in the procedures as written in the protocol.

Vital signs, *Random Blood Glucose* (RBG) and HbA1c levels were examined on day 0 (to serve as baseline data) and 21. The peripheral blood was taken from the cubital vein by the previously trained Public Health Center laboratory analyst. From the collected blood, RBS and HbA1c were measured. Blood pressure, pulse and BMI were checked using tools that have been calibrated by the Calibration and Testing Laboratory of Universitas Ahmad Dahlan. RBS and HbA1c measurements were performed by the Public Health Center laboratory with 5010 spectrophotometer. The brief procedure of HbA1c measurement was as follows: 10 μL of blood plasma was put into the reaction tube, and then added with 1 mL of glucose reagent. After 10 minutes, the patients' blood glucose plasma was measured with a 5010 spectrophotometer. The results were then used to calculate the HbA1c value by the following formula:

$$\mathbf{HbA1c = (Glucose\ Plasma + 77.3) / 35.6}$$

Demographic characteristics were presented descriptively. The mean comparison test was performed on the measurement data of blood pressure, pulse, BMI, serum glucose level, and

IOP Conf. Series: Materials Science and Engineering **259** (2017) 012018 doi:10.1088/1757-899X/259/1/012018

HbA1c levels on day 0 and day 21 with one-way ANOVA using SPSS free edition software (SPSS version 16) with 95% confidence level.

3. Result and Discussion

3.1. *Characteristics of research subjects*

Four hundred and one of diabetes mellitus patients and 615 hypertensive patients who underwent treatment at Jetis I Public Health Center, Bantul, based on the recommendation of the treating physicians, 120 prospective subjects fulfilled the inclusion criteria.

Table 1. Characteristics of research subjects

Characteristics	Treatment Group			p-value
	Placebo group	Group 1	Group 2	
Subjects	n = 33	n = 33	n = 33	
Gender				
Men	9	7	7	0.802
Women	24	26	26	
Age (years)				
≤55 years old	14	14	15	0.582
> 56 years old	19	16	18	
Education				
≤Junior HS	19	21	19	0.999
Senior HS ≤	14	12	14	
Occupation				
Civil servants	1	0	3	0.141
Entrepreneur	5	3	5	
Farmers	4	3	5	
Private Sector	1	2	2	
Labor	5	7	4	
Fishermen	1	0	4	
Unemployed	11	11	5	
etc.	5	7	5	
Marital Status				
Married	32	32	33	0.608
Single	1	1	0	

All prospective subjects were then gathered and briefed and asked to sign the informed consent, but only 112 patients were willing to become research subjects. Demographic and clinical characteristics are presented in table 1 & 2. Table 1 and 2 shows that based on gender, age, education, occupation, marital status and therapy type, there were no differences among groups ($p>0.05$). Most of the subjects were junior high school female graduates and aged over 50 years old.

Most subjects received more than one drug therapy, with most patients receiving a combination therapy of 1 OHO and 1 antihypertensive drugs, followed by patients with a single OHO drug, patients with two OHO drugs, and patients with 2 OHO drugs and 1 antihypertensive medication.

Table 2. Type of drug therapy prescribed by doctors

Type of single and combination therapy for patients at risk of metabolic syndrome	Treatments Group			p-value
	Placebo group	Group 1	Group 2	
	(n = 33)	(n = 33)	(n = 33)	
1 diabetes drug	6	11	6	0.333
1 hypertensive drug	0	0	1	
1 dyslipidemia drug	0	1	0	
2 diabetes drugs	5	6	6	
2 hypertensive drugs	0	1	0	
3 diabetes drugs	0	0	1	
1 diabetes drug +1 hypertensive drug	14	4	7	
1 diabetes drug + 1 dyslipidemia drug	0	1	1	
1 diabetes drug + 2 hypertensive drugs	0	3	1	
2 diabetes drugs + 1 hypertensive drug	5	3	4	
2 diabetes drugs + 1 dyslipidemia drug	1	0	0	
2 diabetes drugs + 2 hypertensive	1	3	3	
3 diabetes drugs + 1 hypertensive drugs	1	0	0	
1 diabetes drug + 1 hypertensive drug + 1 dyslipidemia drug	0	0	1	
1 diabetes drug + 1 hypertensive drug + 2 dyslipidemia drug	0	0	1	
2 diabetes drugs +1 hypertensive drug + 1 dyslipidemia drug	0	0	1	

3.2. *Measurement results of clinical conditions and HbA1c levels on day 0*

Subjects' clinical conditions (blood pressure, pulse rate, BMI, glucose serum level and HbA1c on day 0 (before treatment) are shown in Table 3.

Table 3. Blood pressure, pulse, BMI, serum glucose level and HbA1c on day 0

Vital Signs	Treatment Group			p-value
	Placebo group	Group 1	Group 2	
	(n = 33)	(n = 33)	(n = 33)	
Systolic (mmHg)	142.48± 16.42	141.73±16.84	143.64±18.63	0.903
Diastolic (mmHg)	79.76± 7.005	80.52±9.138	81.09±9.531	0.821
Pulse	89.39± 9.997	89.33±10.764	90.82±11.246	0.816
BMI (Kg/m^2)	24.15± 4.103	23.76±4.346	24.53±3.8	0.749
Serum Glucose Level (mg/dL)	266.93±123.99	233.66±100.94	273.03±126.40	0.070
HbA1c (%)	9,300± 8.057	8.56 ± 2.777	9.34 ± 3.295	0.524

The results of vital signs and HbA1c measurement on day 0 show that there was no significant difference ($p > 0.05$) in blood pressure, pulse, BMI, glucose serum level, and HbA1c levels prior to BCSO administration. Before the treatment, subjects' mean of blood glucose and HbA1c levels exceeded the normal values of RBS with >200 mg/dL and HbA1c with >7%. The mean systolic blood pressure (SBP) also exceeded the normal value with >140 mmHg, but the mean diastolic blood pressure (DBP) was relatively normal with <90 mmHg [8].

3.3. *Measurement results of clinical condition and HbA1c on day 21*

The measurement results of clinical conditions (blood pressure, pulse rate, BMI, blood glucose serum) and HbA1c levels on day 21 are presented in Table 4.

Table 4. Measurement results of clinical conditions and HbA1c levels on day 21

Vital Signs	Treatments Group			p-value
	Placebo group (n = 33)	Group 1 (n = 33)	Group 2 (n = 33)	
Systolic (mmHg)	138.66±13.170	136.10±15.946	139.81±17.662	0.636
Diastolic (mmHg)	75.31±8.476	79.52±17.160	80.87±9.804	0.182
Pulse	89.66±11.085	90.26±10.155	93.87±14.440	0.330
BMI (Kg/m^2)	24.55±4.363	24.04±4.393	25.31±3.881	0.493
Serum glucose level (mg/dL)	270.31±80.324	190.19±88.304	199.61±91.172	0.001
HbA1c (%)	9.60±2.260 [ab]	7.44±2,487 [c]	7.51±2.492	0.001

[a] There was a significant difference between the placebo group and group 1
[b] There was a significant difference between the placebo group and group 2
[c] There was no significant difference between group 1 and group 2.

Table 4 shows that after 20 days of BCSO treatment, there was no difference in blood pressure, pulse rate, and BMI between the treatment groups, with significance value of p>0.05. It shows that the BCSO administration can lower blood pressure to the target level defined for patients with diabetes mellitus by the American Diabetes Association (ADA) and the American Heart Association (AHA) [8], i.e. <140/90 mmHg. However, the result has not met the therapeutic target defined by the International Diabetes Mellitus Federation (IDF) (2006) that recommends a blood pressure target of <135/85 mmHg. There has been a decrease in the mean TDS, i.e. TDS <140 mmHg, although not statistically significant (p> 0.05). Research data also shows that there is no change in average pulse in all groups, and the mean pulse is in accordance with the normal target range according to the American Heart Association (60-100x/min). The average BMI in the three groups after the treatment was relatively similar (p>0.05), categorized as non-obese according to the American Heart Association [8]. The results of this study are not entirely consistent with previous studies, which stated that administering black cumin seed extract (BCSE) with a dose of 500 mg for 8 weeks could lower blood pressure in patients with metabolic syndrome [21]. Differences in terms of intervention timing, intervention type, subjects' initial condition, and research setting contribute to the differences in clinical outcomes. Provided an 8-week BCSE intervention for patients with metabolic syndrome in hospitals, whereas this study provided a 20-day BCSO intervention in patients at risk of metabolic syndrome in Public Health Center [21].

The results showed that there was an average decrease of RBS and HbA1c levels after BCSO treatment for 20 days, and there was a difference in the average values of RBS and HbA1c levels among three treatment groups (p<0.01). There were significant differences in HbA1c and RBS values between the control group and treatment group 1 and treatment group 2 (p<0.01), but there was no significant difference between treatment group 1 and treatment group 2 (p> 0.05). However, despite a decline in the mean HbA1c value of the treatment groups, the results are not yet capable of reaching the targets defined by the American Diabetes Association (<7%). These results are in line with previous studies, which revealed that the administration of BS extract with a dose of 500 mg or BS oil with a dose of 2.5 mL/day for 8 weeks could decrease HbA1c and RBS levels in patients with metabolic syndrome [21] .

BCSO can lower the HbA1c value because it increases insulin production, decreases insulin resistance, stimulates cellular activity, and decreases intestinal insulin absorption [22]. High thymoquinone content in BCSO has antioxidant effect that can decrease oxidative stress to prevent further cell damage [22, 23]. Consumpsion of 2 g/day of black cuminseed combined with oral hypoglycemic drugs can improve the effectiveness of endogenous antioxidant and enzyme therapy [24]. High blood glucose and HbA1c levels have been shown to be associated with increased incidence of death and cardiovascular disease [25]. Furthermore, the antioxidant diet has been shown to inhibit degenerative processes, reduce risk of and death from cardiovascular disease, and prolong life [16, 26-27].

4. Conclusion

Black cumin seed oil treatment with doses of 1.5 and 3 mL/day for 20 days in patients at risk of metabolic syndrome affects HbA1c activity ($p<0.05$) by reducing its level.

Phase III Clinical Trial of BCSO needs to be done in patients at risk of metabolic syndrome with longer administration time as an attempt to obtain phytopharmaca (phytomedicines) for secondary prevention of cardiovascular disease incidence in primary care.

Acknowledgement

The researchers would like to thank the Ministry of Research and Technology and Higher Education for providing the financial aid through Hibah Pascasarjana scheme (no. PPS-058/SP3/III/2016), the research subjects, the staff at Jetis I Public Health Center, Bantul, Yogyakarta, Indonesia and all members of the research team.

References

[1] Aru W S D 2006 *Buku Ajar Ilmu Penyakit Dalam* Jilid III Edisi IV Penerbitan Ilmu Penyakit Dalam FKUI Jakarta Barat.

[2] Leon B M and Maddox T M 2015 Diabetes and cardiovascular disease: Epidemiology, biological mechanisms, treatment recommendations and future research *World J. Diabetes* **6** (13) 1246-1258.

[3] Planiappan L, Mercedes R, Canethon, Yun W, Anthony J G, Hanley, Atephen P F, Stephen M H and Lynne W 2004 Predictors of the incident metabolic syndrome in adults: the insulin resistance atherosclerosis study *Diabetes Care* **3** 788-793.

[4] Kamso S 2007 Metabolic Syndrome in the Indonesian elderly Medical Journal of Indonesia **16** (23) 195-200.

[5] Sudijanto K, Purwantyastuti P, Dharmayanti U L, Ratna J, Yull K R and Besral Besral 2011 Prevalensi dan determinan sindrom metabolik pada kelompok eksekutif di jakarta dan sekitarnya *Jurnal Kesehatan Masyarakat Nasional* **6** (2) 85-90.

[6] Soegondo 2008 Hidup secara mandiri dengan diabetes melitus, kencing manis, sakit gula, FK UI, Jakarta.

[7] Siscovick D S 2015 Type 2 diabetes mellitus and the risk of sudden cardiac arrest in the community *Rev. Endocr. Metab. Disord.* **11** (1) 53–59.

[8] Fox C S, Golden S H, Anderson C, Bray G A, Burke L E, de Boer I H, Deedwania P, Eckel R H, Ershow A G, Fradkin J, Inzucchi S E, Kosiborod M, Nelson R G and Patel M J 2015 Update on prevention of cardiovascular disease in adults with type 2 diabetes mellitus in light of recent evidence: a scientific statement from the american heart association and the american diabetes association *Diabetes Care* **38** 1777–1803.

[9] Khattaba M, Khaderb Y S, Al-Khawaldehd A and Ajlouni K 2010 Factors associated with poor glycemic control among patients with Type 2 diabetes *J. Diabetes Complications* **24** 84–89.

[10] Kayama Y, Raaz U, Jagger A, Adam M, Schellinger I N, Sakamoto M, Suzuki H, Toyama K, Spin J M and Tsao P S 2015 diabetic cardiovascular disease induced by oxidative stress *Int. J. Mol. Sci.* **16** (10) 25234-25263.

[11] Srikanthan K, Feyh A, Visweshwar H, Shapiro J I and Sodhi K 2015 Systematic review of metabolic syndrome biomarkers: a panel for early detection, management, and risk stratification in the West Virginian Population *Int. J. of Med. Sci.* **13** (1) 25-38.

[12] Bajaj S and Khan A 2012 Antioxidants and diabetes *Indian J. Endocrinol. Metab.* **16** Supplement 2 S267-S271.

[13] Kusniyah Y, Nuriswati and Rahayu U 2010 Hubungan tingkat self care dengan tingkat HbA1c pada klien diabetes melitus type 2 di Poliklinik Endokrin RSUP Dr. Hasan Sadikin Bandung.

[14] Zhao D, Yang J and Yang L 2017 Insights for oxidative stress and mtor signaling in myocardial ischemia/reperfusion injury under diabetes *Oxidative Medicine and Cellular Longevity.*

[15] Andrews M, Angone K M, Cray J V, Lewis J A and Johnson P H 1999 Nurses of alternative and complementary therapies.

[16] Zhou M, Zhu L, Cui X, Feng L, Zhao X, He S, Ping F and Li W 2016 Influence of diet on leukocyte telomere length, markers of inflammation and oxidative stress in individuals with varied glucose tolerance: a Chinese population study *Nutrition J.* **15**:39.

[17] Bashandy 2007 Effect of fixed oil *Nigella sativa* on male fertility normal and hyperlipidemic rats B *Int. J. Pharmacol.* **3** (1) 27-33.

[18] Akrom, Darmawan E and Yuhelvia, L 2015 Black cumin seed oilas hepatoprotector in decreasing SGPT and SGOT activity and increasing p53 gene expression in sprague dawley rats induced by alloxan *IJPHS* **4** (3) 159-163.

[19] Akrom and Darmawan E 2017 Tolerability and safety of black cuminseed oil (BSCO) administration for 20 days in healthy subjects *Biomed. Res.* **28** (9) 4196-4201.

[20] Najmi A, S F Haque, M Naseeruddin and R A Kha 2008 Effect of *Nigella sativa* oil on various clinical and biochemical parameters of metabolic syndrome *Int. J. Diabetes & Metab.* **16** 85-87.

[21] Najmi A, Nasiruddin M, Khan R A and Haque S F 2012 Indigenous herbal product *Nigella sativa* proved effective as an antihypetensive in metabolic syndrome **6** (1).

[22] Bamosa and Abdullah O 2015 A Review on the hypoglycemic effect of *Nigella sativa* and thymoquinone *Saudi J. Med. Med. Sci.* **3** (1) 2-7.

[23] Akrom 2013 Mekanisme kemoprefentif ekstrak heksan BJH pada tikus SD diinduksi DMBA: kajian antioksidan dan immunomodulator Disertasi Program Doktor Ilmu Kedokteran dan Kesehatan FK UGM Yogyakarta.

[24] Kaatabi H, Bamosa A O, Badar A, Al-Elq A, Abou-Hozaifa B and Lebda F 2015 *Nigella sativa* improves glycemic control and ameliorates oxidative stress in patients with type 2 diabetes mellitus: placebo controlled participant blinded clinical trial *PLoS ONE* **10** (2) 1-15.

[25] Svensson E, Baggesen L M, Johnsen S P, Pedersen L, Nørrelund H and Buhl E S 2017 Early glycemic control and magnitude of hba1c reduction predict cardiovascular events and mortality: population-based cohort study of 24,752 Metformin Initiators *Diabetes Care* **40** 800–807.

[26] Synder M and Lindsquit 2002 Complementary/alternative therapies in nursing 4th ed, Springer.

[27] Carneiro A and Andrade J P 2017 Nutritional and lifestyle interventions for age-related macular degeneration *A review, oxidative medicine and cellular longevity*, Article ID 6469138, 1-13.

IOP Conf. Series: Materials Science and Engineering **259** (2017) 012019 doi:10.1088/1757-899X/259/1/012019

Effect of calyx capsule-ethanol extract *Hibiscus sabdariffa* L. on renal function of healthy volunteers

N Harun[1,2], E Darmawan[1], L H Nurani[1]

[1]Faculty of Pharmacy, Universitas Ahmad Dahlan, Yogyakarta, Indonesia
[2]Study Program of Pharmacy, STIKes Muhammadiyah Ciamis, Indonesia

E-mail: harunnurhidayati@gmail.com

Abstract. *Hibiscus sabdariffa* contains flavonoid, triterpenoid, anthocyanin which function as immunostimulant. *H. sabdariffa* is considered safe for animal renal; nonetheless, there are known side effects of which need to be further investigated for human renal. This research aims to investigate the effect of calyx capsule-ethanol extract *H. sabdariffa* for renal function of healthy male and female for 30 days period by monitoring Scr and Clcr component in their blood samples. The method of this experimental research was by pre and post-treatment by involving 20 healthy volunteers who have met inclusion and exclusion criteria. The volunteers have completed the informed consent for this experiment. Furthermore, volunteers were divided into two groups (10 male and 10 female). Each group was given orally 500 mg of calyx capsule-ethanol extract *H. sabdariffa* per day for 30 days period. Blood tests were taken on day 0, day 30 after consuming the capsule and day 45 (15 days after the last day of capsule intake) in order to measure the Scr and Clcr concentration in the blood samples by using Jaffe dan Cockcroft-Gault method. The results of each sampling day were further analyzed statistically and compared using Repeated ANOVA dan Friedman test. The results suggest that there was a difference in the renal function on day 0, 30 and 45 samplings. However, there was no significant difference in Scr dan Clcr concentrations on female and male volunteers ($p>0.05$). Specifically, the type of gender affects Scr concentration ($p<0.05$); however, it does not affect Clcr concentration ($p>0.05$). In addition, age and Body Mass Index (BMI) does not affect Scr and Clcr concentrations ($p>0.05$). The side effects discovered through the monitoring increased in mixturition and bloatedness. Calyx capsule-ethanol extract *H. sabdariffa* does not affect on renal function of healthy volunteers.

Keywords: *H. sabdariffa*, healthy renal function, Scr, Clcr

1. Introduction

H. sabdariffa is a natural exogenous antioxidant. It is popular among the well educated women, specifically among the upper middle class who have a great sense of health awareness [1]. *H. sabdariffa* is greatly used to increase stamina since its antioxidant characteristic has immunostimulant effect [2]. The main content of *H. sabdariffa* is vitamin C, flavonoid, polyphenol, beta carotene dan antosianin [3] as well as alkaloid, saponin and tannin [4].

The main component of *H. sabdariffa* i.e. anthocyanin, flavonoid dan chlorogenic acid possess potassium-sparing diuretics, natriuretic in addition to its antioxidant feature [5]. Diuretic is known to maintain the balance of water and electrolyte in the renal system [6].

Renal has a crucial function in the reduction/secretion of medicine/drugs/chemicals in the body [7]. Its function is highly correlated with homeostasis involving urinal, electrolyte, external objects such as toxins and urea secretion. It works through combination of glomerulus filtration (GFR), tubular secretion and body reabsorption [8]. All nutrition and chemicals/drugs will be filtered by GFR hence the GFR capacity is an indicator for renal performance [9].

Glomerulus filtration rate function can be observed from creatinine serum (Scr). Their correlation is not a linear correlation, nevertheless the decrease of GFR is observed when serum creatinine increases, which indicates the reduction in renal cleaning capacity by the reduction of clearence creatinine (Clcr). There is a difference in Scr dan Clcr level of male and female due to difference in Body Mass Index (BMI) [10]. In addition, the difference in eating habits and muscle mass of male and female determine the creatinine production [11].

Drugs effects on renal function are highly correlated with toxic effect which demonstrates the safety of the drugs. Calyx capsul-ethanol extract *H. sabdariffa* is safe for rats with an impact of an increase in Scr depending on the dosage [12]. The impact is found to be toxic for rats once they are given 10,000-66,000 mg/kg body weight of *H. sabdariffa* capsule regularly for a long period [13] also discovers that of *H. sabdariffa* capsule with a dosage of 50,100 dan 200 mg/kg body weight is safe and effective to reduce free radical on rats. In addition, suggests that the LD_{50} (acute toxicity on rats occurs) is 850.90 mg/kg body weight [14]. It is also confirmed that toxicity of *H. sabdariffa* extract is deemed low, as low as LD_{50} 2,000 up to 5,000 mg/kg/day on rats [15].

Based on the findings above, there is a need to further study the effect of calyx capsule-ethanol extract *H. sabdariffa* on renal function of healthy male and female.

2. Materials and Methods

This research was conducted using pre and post treatment to determine the effects of *H. sabdariffa* on renal function of 20 healthy volunteers that divided into two groups (10 male and 10 female).

Inclusion criteria in this research were age > 18-45 years, not smoking, IBM 18-30 kg/m^2, not taking supplements within 7 days before and at the time of treatment who have a health certificate from a registered doctor. While exclusion criteria were pregnant and breastfeeding, using hormonal contraceptives, autoimmune disease, and history of renal impairment.

2.1. Variable

The dependent variables in this research were measure the Scr and Clcr concentration in the blood samples on day 0, day 30 and day 45. The independent variables in this research were 500 mg calyx capsule-ethanol extract *H. sabdariffa* orally per day for 30 days period for all volunteers. The confounding variables were body habitus and diet control.

2.2. Place and time research

This research was conducted in Hidayatullah Hospital of Yogyakarta for 45 days (April 3rd to May 17th 2017).

2.3. Instruments

Serum creatinine measured by applying Jaffe method using spectrophotometer (486 nm wave length) and picric acid as a reactor. In addition, the measurement of Clcr is performed by Cockcroft-Gault equation.

2.4. Health check and preparation

The potential volunteers who have filled the informed consent are physically examined for their hematology, liver and renal by registered doctors. The results are recorded in physical examination and laboratories reports.

The 20 volunteers who have been approved by the Ethical Committee of Universitas Muhammadiyah Yogyakarta (Approval Number No. 255/EP-FKIK-UMY/IV/2017). The volunteers were given orally of 500 mg calyx capsule-ethanol extract *Hibiscus sabdariffa* (product code/batch number: 5055C/RH162703) per day after meals for 30 days period.

Blood samples are taken on median cubital vein or brachial vein using 5 cc syringe for 4 cc blood then put into microsentrifugation tube containing EDTA as anticoagulant. Plasma separation carried out by centrifugation with a speed of 3500 rpm for 2 hours resulting a clear yellow liquid (serum). If the serum not used immediately, it is stored in the freezer with a temperature of -20^{0}C. 0.25 cc to 0.5 cc serum is taken for Scr measurement. Serum that reacts with picric acid placed inside auto-analyzer photometric to measure Scr by applying Jaffe method. Once Scr concentration obtained, Clcr estimated by Cockcroft-Gault equation as follows:

$$\text{Male : Clcr} = (140 - \text{age})\ \text{ABW}/(\text{Scr} \times 72)$$
$$\text{Female: Clcr} \times 0.85$$

2.5. Data analysis
The collected data were tabulated. Continuous variables were presented as mean±SD. We used SPSS 23.0 for data analysis. Scr measurement and Clcr estimation are analyzed using Repeated ANOVA and Friedman test with 95% confidence interval.

3. Results and Discussion
3.1. Scr and Clcr Measurements
The results of Scr measurement and Clcr estimation from the volunteers on day 0, 30 and 45 (15 days after the last day of capsule intake) is shown in Table 1 dan 2.

Table 1. Average male (n=10) renal function on day 0, 30 and 45 (15 days after the last day of capsule intake

Parameter	(Mean±SD) Day 0	Day 30	Day 45	P value	Normal Value
Scr (mg/dL)	1.07±0.31	1.14±0.12	1.22±0.16	0.050	0.5-1.2
Clcr (mL/min)	93.05±29.45	82.99±18.94	77.93±18.64	0.193	80–130

Table 2. Average Female (n=10) Renal Function on day 0, 30 and 45 (15 days after the last day of capsule intake

Parameter	(Mean±SD) Day 0	Day 30	Day 45	P value	Normal Value
Scr (mg/dL)	0.84±0.22	0.96±0.11	0.96±0.14	0.132	0.4-1.1
Clcr (mL/min)	85.08±18.19	71.70±8.02	72.93±15.20	0.181	75-120

Notes:
day 0: before *hibiscus sabdariffa* capsule intake
day 30: after *hibiscus sabdariffa* capsule intake (last day)
day 45: 15 days after the last day of capsule intake

Table 1 shows that there is no difference in male renal function on day 0, day 30 and day 45 based on Scr and Clcr content (p>0,05). Likewise, Table 2 demonstrate the same result for female renal function.

Both female and male renal function (Scr) is not affected by the *H. sabdariffa* capsule intake from day 0, day 30 as well as day 45 (p>0.05). However, there is a difference in the average Scr level on day 0 and day 30 of which Scr increases on day 30 within the acceptable level for human renal. Toxic effect occurs at 10,000-66,000 mg/kg body weight *H. sabdariffa* intake for rats (regular and long period intake). The effects could be muscular dystrophy leading to renal disturbance [12]. The increase in Scr level is due to the content of anthocyanin and quersetin found in *H. sabdariffa* extract. These flavonoid substances are effective free radicals scavenger [16]. Anthocyanin found in the extract ethanol is less than one using water [12] hence the increase of Scr is not found to be significant in this research. If muscular dystrophy occurs, the creatinine production will diminish due to muscle formation inefficiency leading to increase in creatinine in the blood [17]. The Muscular dystrophy effect is not discovered from the volunteers since the creatine in the blood is still within permissible 50-100 mg/kg body weight.

Scr measurement on male volunteers shown that there is a surge of Scr on day 45 which shows that there is a potential *H. sabdariffa* residue or other influencing factors such as low water intake or unhealthy lifestyle. In contrast, there is no surge in Scr on day 45 on female volunteers.

There is an insignificant difference in Clcr estimations for male and female volunteer ($p>0.05$). There is a decrease in Clcr level on day 30 for male and female volunteers; however it is still within permissible limit. The reduction in Clcr is due to the increase in creatinine production being wasted by body [18]. The increase in creatinine by products is due to quersetin content which stimulates the release of Nitric Oxide (NO) on vascular endothelial. This phenomenon is known as vasorelaxation which results in failure in renal function [5].

3.2. *Volunteers response to Scr and Clcr monitoring*

Creatinine level does not only depend on muscle mass but also muscle activities, diet, health conditions, age, gender, and race [19]. Scr and Clcr monitoring based on gender, body mass index (BMI) and age is provided in Table 3.

Table 3. Scr and Clcr monitoring based on gender, body mass index (BMI) and age.

Group	Scr (Mean±SD)	p-value	Clcr (Mean±SD)	p-value
Gender				
Male	1.14±0.11	0.001	82.99±18.94	0.207
Female	0.96±0.11		72.94±15.21	
Body Mass Index				
<18.5	1.2±0.00	0.495	64.95±0.00	0.262
18.5-22.9	1.03±0.13		74.78±16.96	
23.0-24.9	1.03±0.21		92.31±16.32	
25.0-29.9	1.2±0.00		95.70±0.00	
Age	1.05±1.39	0.621	77.79±17.49	0.559

Clcr level is not affected by age, BMI and age ($p>0.05$); nonetheless, Scr is affected gender factor ($p<0.05$) due to difference is muscle mass. Apart from muscle mass, the difference in Scr is due to the difference in muscle mass and diet in different gender [8].

3.3. *The side effects*

The side effects discovered from intake of calyx capsule-ethanol extract *H. sabdariffa* are increase in micturition and bloatedness. It is found that 3 male experiences micturition increase and 2 male experiences bloatedness. Increase in micturition can still be considered as minor diuretic. This result is aligned with the result from rats [20]. It is found that *H. sabdariffa* can increase renal filtration by 48% since quersetin found in *H. sabdariffa* stimulates the release of Nitric Oxide (NO) on vascular endothelial. This phenomenon is known as vasorelaxation which results in diuretic effect. Diuretic effect can be found in the volunteers who consume the *H. sabdariffa* with dosage of 25, 50, and 100 mg/ body weight [21].

Bloatedness occurs since calyx capsul-ethanol extract *H. sabdariffa* contains ascorbic acid (260-280 mg/100 g) found in the dray leaves of hydroxycitric acid and hibiscus acid [22]. Bloatedness can results in discomfort in the digestion system.

3.4. Research limitation
Research limitation were variables measured only in adult volunteers, without measuring the elderly and children volunteers. As well this research only measured sub chronic toxicity, further research is needed to determine chronic toxicity.

4. Conclusions
It can be concluded that calyx capsule-ethanol extract *H. sabdariffa* does not affect on renal function of healthy volunteers.

Acknowledgement
The authors thank to volunteers who have been willing to participate in this research and to Hidayatullah Hospital who has assisted in the process of volunteer examination.

References
[1] Muhaimin A W 2010 Perilaku konsumen dalam penelitian teh rosella merah di Kota Malang *Agritek* (2)18.
[2] Ekanto B, Sugiarto 2011 Kajian teh rosella (*Hibiscus sabdariffa*) dalam meningkatkan kemampuan fisik berenang (penelitian eksperimen pada mencit jantan remaja) *J Media Ilmu Keolahragaan Indonesia* **1**(2) 171-180.
[3] Pangkahlila W P 2007 *Memperlambat penuaan meningkatkan kualitas hidup anti-aging medicine* (Jakarta: Penerbit buku Kompas) 8-11.
[4] Okereke C N, Iroka F C, Chukwuma M O 2015 Phytochemical analysis and medicinal uses of *Hibiscus sabdariffa Int J Herb Med* **2**(6) 16-19.
[5] Jiménez F E, Alarcón-Alonso Javier, Aguilar-Rojas Arturo 2012 Diuretic effect of compounds from *hibiscus sabdariffa* by modulation of the aldosterone activity *Planta Medica* **78**(18) 1893-8.
[6] Davey P 2010 *Medicine at a Glance* 3rd edition (Oxfort: Blackwell's publishing) p 31 .
[7] Neal JM 2012 *Medical Pharmacology at a Glance* 7 Edition (Oxford: John Wiley & Sons Ltd.) 12-13.
[8] DiPiro J T P D, Talbert R L, Yee G R, Matzke G R P, Wells B G and Posey L M 2008 *Pharmacotherapy: A Pathophysiologic Approach* (New York: McGraw-Hill) 49-50.
[9] Bauer L A 2008 *Applied Clinical Pharmacokinetics* second edition (New York: McGraw-Hill) 51-53.
[10] Guyton A C M, Hall J E P 2006 *Text Book of Medical Physiology* 8 edition (Philadelphia: Elsevier Saunder) chapter 27 344-347.
[11] Groesbeck D, Köttgen A, Parekh R, Selvin E, Schwartz G J, Coresh J F S 2008 Age, gender, and race effects on cystatin C levels in US adolescents *Clint J Am Soc Nephrol* **3** (6) 1777-1785.
[12] Fakeye T O, Pal A, Bawankule D U, Yadav N P, Khanuja S P S 2009 Toxic effects of oral administration of extracts of dried calyx of *Hibiscus sabdariffa* linn (*malvaceae*) *Phyther Res.* **23** (3)2664.
[13] Nurkhasanah, Bachri M S, Aziz N R 2016 Pengaruh pemberian subkronik ekstrak kelopak bunga rosella (*Hibiscus Sabdariffa* L.) terhadap kadar SGPT, SGOT, dan ALP *J. Farm. Sains dan Komunitas* **13** (2) 95.
[14] Fita Sari, Nurkhasanah M 2016 Uji toksisitas akut extract etanol kelopak rosella (*Hibiscus sabdariffa* L.) pada tikus sprague dawley *Trad. Med. J.* **21** (1) 17.
[15] Hopkins A L, Lamm M G, Funk J L 2013 *Hibiscus sabdariffa* L. in the treatment of hypertension and hyperlipidemia: a comprehensive review of animal and human studies

Fitoterapia **85** (1) 87
[16] Choe E, Min D B 2009 Mechanisms of antioxidants in the oxidation of foods *comprehensive reviews in food science and food safety* **8** (1) 353.
[17] Zhang H, Zhu Y, Sun Y, Liang Y, Li Y, Zhang Y 2015 Serum creatinine level: A supplemental index to distinguish Duchenne muscular dystrophy from Becker muscular dystrophy *Dis Markers* **2015** ((141856) 5.
[18] Lascano M E, Poggio E D 2013 Kidney function assessment by creatinine-based estimation equations *Dis Manag Proj Nephrology* (2010*)*1-3.
[19] Alarcón-Alonso J, Zamilpa A, Aguilar F A, Herrera Ruiz M, Tortoriello J, Jimenez Ferrer E 2012 Pharmacological characterization of the diuretic effect of *Hibiscus sabdariffa* Linn (*Malvaceae)* extract *J. Ethnopharmacol.* **xxx** (2011) 5.
[20] Yulinah E, Wahyuningsih S, Ratna K 2015 Efek diuretik ekstrak air kelopak bunga rosela (*Hibiscus sabdariffa* Linn.) pada tikus wistar jantan *J Farm SAINS dan Terapan.* **2** (2) 4.
[21] Da Costa Rocha I, Bonnlaender B, Sievers H, Pischel I, Heinrich M 2014 *Hibiscus sabdariffa* L.- A phytochemical and pharmacological review *Food Chemistry.* **165** (2014) 427.

The ability of *Abelmoschus manihot* L. leaf extract in scavenging of free radical DPPH and total flavonoid determination

S Sudewi[1], W A Lolo[1], M Warongan[1], Y Rifai[2], H Rante[2]

[1] Department of Pharmacy FMIPA, Universitas Sam Ratulangi Manado, Indonesia

[2] Faculty of Pharmacy, Universitas Hasanudin Makassar, Indonesia

Email: ssudewi@gmail.com

Abstract. *Abelmoschus manihot* L. has reported to have flavonoids content. This study aims were to determine the ability of *A. manihot* extract in counteracting free radical DPPH and determine the content of total flavonoids. *A. manihot* leaf was taken from 2 regions in North Sulawesi, namely Tomohon and Kotamobagu. The maceration was carried out to extract the active compound in a 96% ethanol solvent. Free radical scavenging analysis was carried out by DPPH and determination of its total flavonoid in the extract was measured using spectrophotometri method. The results showed that *A. manihot* extract from Tomohon and Kotamobagu could counteract free radical of DPPH with value of free radical activity of 88.151 and 88.801 %, respectively. *A. manihot* leaf from Kotamobagu has higher total flavonoids content 61.763 mg/g compare to Tomohon 46.679 mg/g which presented as quercetin. *A. manihot* has antioxidant activity.

Keywords: *Abelmoschus manihot* L. leaf extract, free radical DPPH, total flavonoids

1. Introduction

In North Sulawesi, *Abelmoschus manihot* L. or known by the name of the Gedi is one of the plants of the common Malvaceae tribe planted. Easily grown stem cuttings on loose soil. The society recognizes two types of *A. manihot* ie red Gedi and green Gedi. Green leaves can be utilized by the Manado people as foodstuffs known as Manado Porridge (Tinutuan), typical of Manado, whereas for traditional red Gedi can cure some diseases [1].

The method which could be used to measure antioxidant capacity is the DPPH method (1,1-diphenyl-2-picrylhydrazyl). DPPH is widely used to test an antioxidant activity as a scavenger of free radicals of food and herbal extract. DPPH can also be used to quantify antioxidants in complex biological systems in recent years. The DPPH method can be used in both solid and liquid samples and could be applied to the whole antioxidant capacity present in the sample.

IOP Conf. Series: Materials Science and Engineering **259** (2017) 012020 doi:10.1088/1757-899X/259/1/012020

It has been extensively studied about the benefits of *A. manihot* including antioxidant activity [2], analgesics [3] as well as formulations [4]. Previous study reported antioxidant activity of the *A. Manihot* leaves which found from the areas of Palu, Makassar and Gorontalo [5]. The identification of chemical compounds reported the presence of flavonoid in *A. manihot* leaves extracted with ethanol solvent [6]. The level of flavonoid content of A. Manihot leaves extracted with 96% ethanol was 41.56% [5]. Flavonoid compounds have various important functions for health, among others, in reducing the risk of cardiovascular disease, blood pressure, atherosclerosis, and as an antioxidant [7]. This study aims to determine the ability of *A. manihot* extract in counteracting free radical compounds DPPH and determine the total content of flavonoids taken from 2 regions in North Sulawesi, namely Tomohon and Kotamobagu.

2. Materials and Methods

2.1 Plant material

A. manihot leaf samples were taken from two locations in North Sulawesi: Tomohon and Kotamobagu. The sample is washed with clean running water. The next process is cut into small pieces and dried. After drying, the sample was powdered and then made stage of extract making.

2.2. Extraction of A. manihot leaves

A. manihot leaf powder was macerated using a 96% ethanol solvent of 500 mL. Remacerated was carried out with 300 mL of 96% ethanol. The macerate was then filtered. The macerat was evaporated with a rotary evaporator until a viscous extract was obtained. Further dried in the oven to obtain a dried extract.

2.3 Free radical scavenger Activity with DPPH Method

This test was conducted to determine the presence or absence of antioxidant activity as a free radical scavenger [8]. A total of 1 mL of *A. manihot* extract with concentration of 150 µg / mL was added with 2 mL of DPPH solution in 0.08 mM methanol. The mixture is then divorted and left for 30 minutes at room temperature in the dark. The absorbance was measured at a wavelength of 517 nm and methanol was used as a blank. The percentage of DPPH free radical scavenger activity is calculated according to the equation:

$$\text{Free radical scavenger activity (\%)} = \frac{\text{Abs.control} - \text{Abs.Sample}}{\text{Abs.control}} \text{ x } 100\% \qquad (1)$$

2.4 Determination of total flavonoids

A sample of 0.1 mL of *A. manihot* leaf extract was put into a test tube, added 3 ml of Aluminum chloride solution then mix by vortex for 1 min. The mixture was incubated for 30 minutes. The absorbance of the solution was determined using a Spectrophotometer at a wavelength of 415 nm. The presence of flavonoids is shown by the formation of colors [9].

2.5 Characterization of functional group by FTIR

A small quantity *A. manihot* extract were mixed with KBr. The functional group were analysed by Fourier Transform Infra Red Spectrophotometer (Shimadzu) in region of 4000-400 cm^{-1}.

3. Results and Discussion

3.1 Extraction

Extraction is done to draw chemical components contained in a material. Maceration is choosen because this method is done by soaking the sample powder without heating so thermal decomposition can be avoided. The viscous extract obtained was dark green with the percentage of rendement in the Table 1.

Table 1. Rendemen of *A. manihot* extract

Name	Rendemen (%)
Tomohon	2,366
Kotamobagu	5,796

3.2 Free radical scavenging activity
The antioxidant activity in this research was evaluated using DPPH method. The DPPH was chosen because this method is a simple, easy, fast and sensitive method and requires only a small sample to know the antioxidant activity of natural compounds [10]. Before testing of the extract, determination of the maximum wavelength and operating time was carried out. The results of the determination of the maximum wavelength in this study was 517 nm and the operating time was 30 minutes for the incubation time. The results of free radical scavenging activity testing can be seen in the Table 2.

Table 2. Free Radical scavenger activity of *A. manihot* leaf of Tomohon

Concentration (µg/mL)	Absorbance			% Free radical inhibitory
	I	II	Average absorbance	
50	0.233	0.164	0.1985	81.552
100	0.206	0.142	0.174	83.829
150	0.189	0.103	0.146	86.431
200	0.173	0.109	0.141	86.896
250	0.157	0.098	0.1275	88.151

Table 3. Radical scavenging activity of *A. manihot* leaf from Kotamobagu

Consentration (µg/mL)	Absorbance			% free radical inhibitory
	I	II	Average absorbance	
50	0.238	0.253	0.2455	77.184
100	0.217	0.164	0.1905	82.296
150	0.186	0.158	0.172	84.015
200	0.162	0.171	0.1665	84.526
250	0.103	0.138	0.1205	88.801

The existence of antioxidant activity could be caused by the compound contained in plants release the H atom and then bind to the radical DPPH form a new compound which more stable was diphenyl picrilhidrazine [11].

3.3 Total flavonoids
Determination of total flavonoid from *A. manihot* leaf extract was done according to previous [9]. Before the total measurement of flavonoids in the sample, the maximum wavelength was determined first. Maximum wavelength measurement results were obtained at a wavelength of 417 nm. The operating time stability was determined and obtained at 30 minutes. The total flavonoid measurements were determined at a wavelength of 417 nm and 30 min incubation time [12].The principle of total measurement of flavonoids based on color formation with $AlCl_3$ reagents. Flavonoid has a keto group on C-4 and has a hydroxy group on C-3 or C-5 atoms which can form a color complex with $AlCl_3$

IOP Conf. Series: Materials Science and Engineering **259** (2017) 012020 doi:10.1088/1757-899X/259/1/012020

[13]. The addition of acetic acid was to maintain the complex of C-4 keto and 3 or 5 -OH remains stable. The result of determination of quercetin standard curve can be seen the Figure 1.

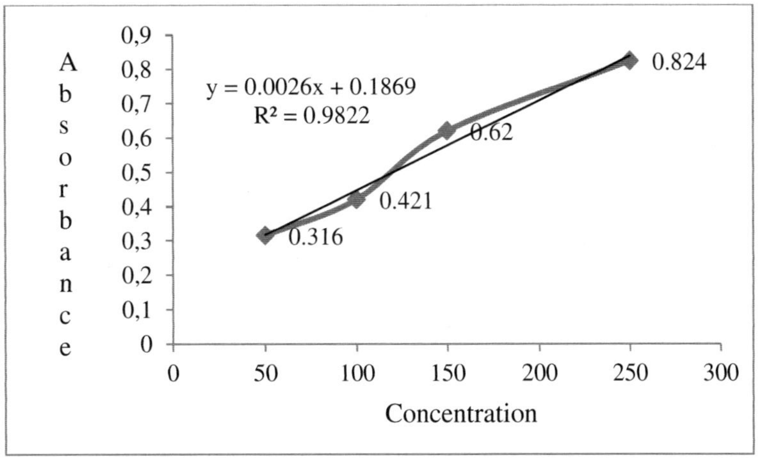

Figure 1 . Standard curve of quercetine for total flavonoid

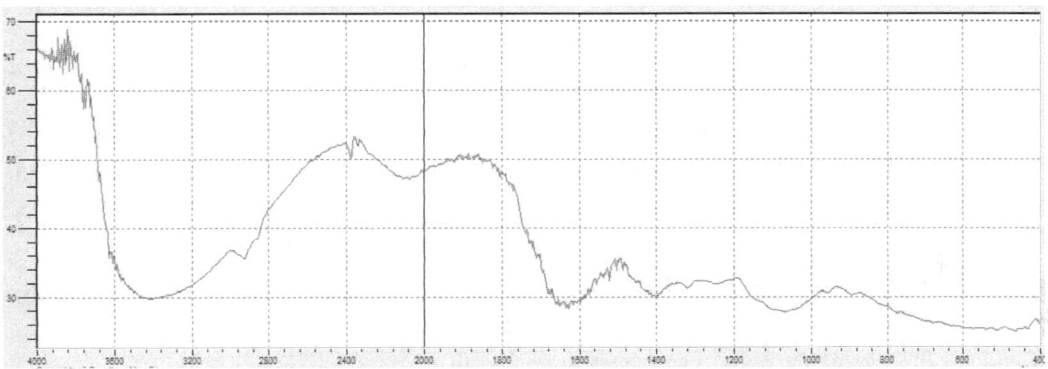

Figure 2 . FTIR spectra of *A. manihot* from Kotamobagu

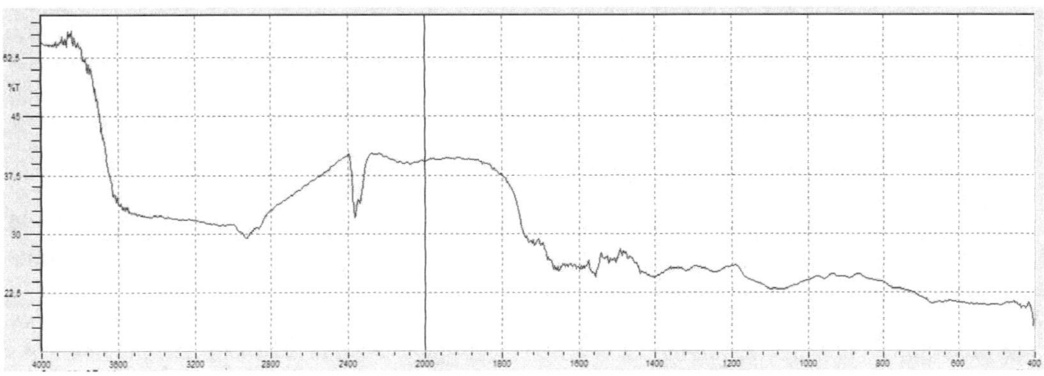

Figure 3 . FTIR spectra of *A. manihot* fromTomohon

The *A. manihot* leaves from Tomohon and Kotamobagu towns are estimated to have functional groups -OH as shown in the Figure 2 and figure 3. The absorption in wave numbers 3000-3300 cm^{-1} is typical hydroxyl group [14].

4. Conclusions

A. manihot leaf extract from Tomohon and Kotamobagu is able to scavenge of DPPH free radical.

Aknowledgment
The authors thanks to Ministry of Research, Technology and Higher Education for research funding through Pekerti Scheme year 2017/2018.

References

[1] Mamahit L P and Soekarno N H 2010 Satu senyawa organik yang diisolasi dari daun Gedi (*Abelmoschus manihot* L) asal Sulawesi Utara *Chemistry Progress* **3(1)** 45.
[2] Mercy T, Sri R, Pudji H and Agnes M 2015 Ekstraksi daun Gedi (abelmoschus manihot l) secara sekuensial dan aktivitas antioksidannya *Agritech.* **35 (3)** 280-287.
[3] Ristanti P, Jimmy P and Fatimawali 2013 Uji efek analgesik ekstrak etanol daun Gedi (*Abelmoschus manihot* (L.) Medik) pada mencit (*Mus musculus*) *Jurnal e-Biomedik* **1(1)** 571-580.
[4] Winda M R, Paulina V Y, Yamlean and Sudewi S 2016 Formulasi dan evaluasi sediaan tablet ekstrak daun Gedi hijau (*Abelmoschus Manihot*) dengan metode granulasi basah. *Pharmacon* **5(2)** 243-250.
[5] Pine, A T D, Alam G and Attamin F 2011 Standarisasi mutu ekstrak daun Gedi (*Abelmoschus manihot*) dan uji efek antioksidan dengan metode DPPH. *e-journal pascasarjana UNHAS.* Makassar.
[6] Jeni, T 1992 *Pemeriksaan Kandungan Kimia Daun Gedi (Abelmoschus manihot L. Medik).* Fakultas Matematika dan Ilmu Pengetahuan Alam. Institut Teknologi Bandung, Bandung.
[7] Hodgson J M and Kevin D C 2006 Review dietary flavonoids: Effects on endothelial function and blood pressure *Journal Science Food Agriculture* **86** 2492-2498.
[8] Chandini S K, Ganesan P and Bhaskar N 2008 In vitro antioxidant activities of there selected brown seaweeds of India *Food Chemistry* **107(2)** 707-713.
[9] Meda A, C E Lamien, M Romito, J Milliogo and O G Nacoulina 2005 Determination of the total phenolic flavonoid, and proline content in Burkina fasan money, as well as their radical scavenging activity. *Food Chemistry* **91** 571-577.
[10] Koleva I I, Van Beek T A, Linssen J P H, Groot A De and Evstatieva L N 2002 Screening of plant extracts for antioxidant activity: A comparative study on three testing methods. *Phytochemical Analysis* **13** 8-17.
[11] Molyneux R 2004 The use of stable free radical diphenylpicrylhydrazyl (DPPH) for estimating antioxidant activity *Songklanakarin Journal of Sciences* **2** 211-219.
[12] Lamien-Meda A, Lamien C, Compaoré M, Meda R, Kiendre-beogo M, Zeba B, Millogo J and Nacoulma O 2008 Polyphenol content and antioxidant activity of four-teen wild edible fruits from Burkina Faso Molecules **13** 581-594.
[13] Desmiaty Y, Ratnawati J and Andini P 2009 *Penentuan Jumlah Flavonoid Total Ekstrak Etanol Daun Buah Merah (Pandanus conoideus* Lamk.*) Secara Kolorimetri Komplementer.* Dipresentasikan Pada Seminar Nasional POKJANAS TOI XXXVI 13 & 14 Mei 2009, Universitas Sanata Dharma Yogyakarta.
[14] Silverstein 2002 *Identification of Organic Compund,* 3rd Edition (Neew York: John Wiley & Sons Ltd).

IOP Conf. Series: Materials Science and Engineering **259** (2017) 012021 doi:10.1088/1757-899X/259/1/012021

Utilization of a hospital information system for outpatient prescription screening process at the PKU Muhammadiyah Yogyakarta Hospital

R Ismail[1], D A Perwitasari[1], W Supadmi[1], I Risdiana[2]

[1]Faculty of Pharmacy, Universitas Ahmad Dahlan, Yogyakarta, Indonesia

[2]PKU Muhammadiyah Yogyakarta Hospital, Indonesia

E-mail: rahmatismail114@gmail.com

Abstract. Prescription screening includes administrative and clinical precision of the drug, dosage, frequency and route of administration, therapeutic duplication, allergic or sensitive reactions, and actual or potential interactions. The study was aimed to identify the obstacles and compliance level of users, as well as the design of a prescription screening information system and its users' perceptions. This study used qualitative and quantitative research design with action research studies involving pharmacists, pharmacy technicians, a programmer and clinical practice student pharmacists. The obstacle of pharmacists in doing prescription screening was the long duration in the process of manual prescription review. The compliance of pharmacists in manual prescription review was under 50%. The prescription information system was created by the programmer on the proposal of pharmacists in the form of front view, pharmacy display, sales display, prescription display, prescription display per period and display of recapped prescriptions. Perception of the usefulness was very high with a value of 4.5±0.577 and perception of ease of use was very high with a value of 4.214±0.534 from 28 respondents. The prescription information system was created by a programmer upon the recommendations of pharmacists. Perception of the usefulness and ease of use was very high.

Keywords: prescription screening, prescription information system, user perception

1. Introduction

The use of an electronic information system is a highly important support in improving the quality of hospital services. More specifically, it is utilized in pharmaceutical installations for medication management. Survey has shown that a computer-based information system management can cut total expenses by up to 5% [1].

The process of prescription screening and drug evaluation is performed by a pharmacist on: (a) The accuracy of the drug, dose, frequency and route of administration; (b) Therapeutic duplication; (c) Actual or potential allergy or sensitive reactions; (d) Actual or potential interaction between the medicine and other drugs or food; (e) Variation in the administration criteria designated by the hospital; (f) The patient's body weight and other physiological information; (g) Other contraindications [2]. The prescription screening process is part of the application of pharmaceutical care [3].

Prescription review is still carried out manually, from recording into prescription screening files, registry, until the administration and presentation of report data. Moreover, interviews with prescription screening clerks have revealed that the process prolongs their work, to the extent that clerks face difficulties in recapitulating prescription review data for administration at the end of each month. A manual reporting process potentially generates errors and requires longer screening time [4].

A prescription screening information system is thus required to enhance prompt and accurate prescription screening services. Prescription screening results are then reported to the pharmacy and therapeutics committee to be analyzed and given corrective suggestions. In addition, doctors must also receive inputs about prescribing patterns with the potential for frequent errors [5, 6].

2. Materials and Methods

2.1. Research design

This was a qualitative and quantitative study with an action research design, which included a research and a participation stage involving pharmacists, pharmacy technicians, a programmer and student pharmacists in clinical internship, as well as an action stage which aimed to alter the condition of a group of people.

2.2. Research location

This study was conducted at the PKU Muhammadiyah Yogyakarta Hospital because a clinical information system development and a pharmaceutical care program were being implemented at this location.

2.3. Research subjects

The subjects consisted of 41 individuals, comprising 1 decision maker for the information system, 1 programmer, 8 pharmacy practitioners, 9 pharmacy technicians, and 22 student pharmacists practicing at the PKU Muhammadiyah Yogyakarta Hospital.

2.4. Research instruments

The instruments utilized for this study were (1) focus group discussion guidelines, (2) group interview guidelines, and (3) a questionnaire.

2.5. Research stages and data collection methods

2.5.1. Diagnosing stage

Researchers reviewed prescription screening done manually by the pharmacists covering 997 prescriptions in May, 1,340 in June and 1,239 in July from four days of prescription services per month in 2015.

2.5.2. Planning stage

A group interview was held twice; the first was attended by the 8 pharmacy practitioners and the programmer, while the second was attended by 3 pharmacy practitioners, the person in charge (PIC) of the hospital information system and the programmer.

2.5.3. Action stage

In this stage the programmer built a prescription screening information system based on the recommendations of the pharmacy practitioners.

2.5.4. Evaluation stage

This stage employed a questionnaire survey method to identify the perceptions of its users.

2.6. Data analysis

Analysis of the data in this study was run by means of content and descriptive analyses. Content analysis was undertaken to examine the results of the manual prescription screening, the focus group discussion and the group interview. Descriptive analysis was applied to assess the utility and ease of access of the information provided by the prescription screening information system by calculating the percentage of each question or statement over the respondents' answers.

3. Results and Discussions

3.1. Description of manual prescription screening

In this study prescription screening was practiced manually by collecting prescriptions from May, June and July 2015 to oversee the compliance of the staff in following the procedure. According to the previous study, the process of prescription screening includes administrative terms, pharmaceutical suitability and clinical considerations. The latter two comprise preparation form, dose, potency, stability, incompatibility, route and period of administration, presence of allergy, side effects, interaction, and suitability (dose, duration, drug amount, etc.) [7]. The adherence of the PKU Muhammadiyah Yogyakarta Hospital pharmaceutical personnel in performing prescription screening was considerably low, as can be seen in Figure 1.

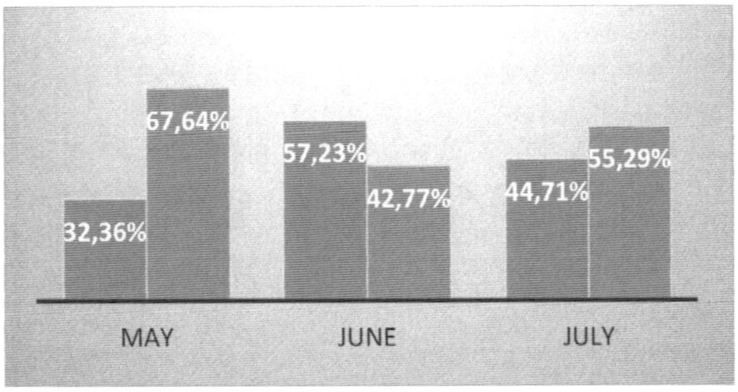

Figure 1. Percentages of personal compliance

According to Figure 1, the observance of the outpatient pharmaceutical staff at PKU Muhammadiyah Yogyakarta Hospital in screening a total of 3,576 prescriptions was 32.36% of 998 in May, 57.23% of 1,341 in June and 44.71% of 1,239 in July. Overall, the compliance level of the staff in conducting prescription screening was below 50%.

3.2. Prescription screening obstacles

Table 1. FGD results on prescription screening obstacles at the pharmaceutical installation

FGD Participant	Result
1	Not much time for the prescription screening process
2	Prescription screening form filled out only when there is an issue with the prescription
3	Uncertain about the filled prescription screening form because of literature limitations
4	Filling out the form within available time while considering the queue of patients collecting their medications

From results of the FGD as presented in Table 1, it can be concluded that the hurdles faced by the staff in operating the pharmaceutical installation prescription review process were caused by: (a) Lengthy use of time in the manual prescription screening process; (b) lack of confidence in performing prescription screening due to limited literature; (c) filling out the prescription screening form only if there is an issue in the prescription; and (d) the abundance of drug prescriptions to review.

3.3. Prescription screening information system needs analysis

Table 2. FGD results on prescription screening information system needs

FGD Participant	Result
1	Procuring prescription screening software by exploiting the hospital information system
2	Utilizing software which automatically reviews prescriptions in order to minimize the time required for manual prescription screening
3	Training prescription screening operators

Table 2 shows that a prescription screening information system is strongly necessary to minimize errors and reduce the time needed for the process. Prescription screening by means of an information system boasts a high degree of confidence with the integration of reliable literature into the prescription screening software. The system itself employs a computer (hardware/software), a procedure manual, a model of analysis, planning, control and decision-making, and a database [8].

Following up recommendations from pharmacy practitioners and technicians to apply a prescription screening information system, the first group interview was organized between the practitioners, the programmer and the PIC of the PKU Muhammadiyah Yogyakarta Hospital information system.

Table 3. First group interview results on suggestions for a prescription screening information system design

Problem	Description
Evaluation of the prescription screening information system by attending peer pharmacists	Certain clinical pharmacy activities need the help of computer technology to accelerate activity reporting and documentation, including prescription review, to enable the presentation of prescription screening documents to doctors or the Medical Committee as material for the evaluation of prescribing issues.
Prescription screening form design	The design adhered to the standard guidelines of accreditation.
Drug information master data as prescription screening support	The purchase of Lexicomp® or Medscape® software would be proposed to fulfil the need for supporting data.

It can be noted from Table 3 that the group interview held to conceive the prescription screening software bore outcomes as follows: (a) The prescription screening form from the accreditation standards would be adapted to the requirements of the PKU Hospital and refer to other sources; (b) the prescription screening information system program would be integrated with the other hospital information system programs; (c) the prescription screening information system program would be installed at the outpatient and inpatient pharmacy, except in the patient prescription billing computer to avoid inhibiting the prescription service rate; and (d) a clinical drug information master data would be devised to support the prescription screening process by using internal and external sources, namely the drug master data recapitulated in the pharmaceutical management information system, and by acquiring paid software (Medscape®/Lexicomp®).

After the first, a second group interview was held with the pharmacy practitioners and programmer as participants.

Table 4. Second group interview results on perfecting the prescription screening information system

Problem	Description
Programmer's presentation of a prescription screening information system sample	The prescription screening information system made by the programmer underwent a preliminary simulation.
Additional prescription screening items proposed by pharmacy practitioners	The software created by the programmer needed revision and addition of several items, such as repairing the "click/entry" menu and inserting a BPJS intervention item.

As in Table 4, results of the second group interview decided to refine the prescription screening information system by including BPJS intervention and insurance suitability items to accommodate changes of drug administration in cases of insurance/BPJS-covered patients. The prescription screening information system was accordingly improved by the programmer.

IOP Conf. Series: Materials Science and Engineering **259** (2017) 012021 doi:10.1088/1757-899X/259/1/012021

3.4. Prescription screening information system design

Herewith is the final iteration of the prescription screening information system built by the programmer based on inputs from the pharmacists:

1. Front Display: Registration, billing, medical, support, logistics, pharmacy, medical records, nursing care, setting and windows.
2. Pharmacy Menu Display: [FAR-01] prescription, [FAR-02] prescription list, [FAR-20] UDD, [FAR-21] prescription queue list, [FAR-03] sales/DO-bill (DU-General), [FAR-09] sales/DO-bill (DU-type), [FAR-04] item handover, [FAR-05] outstanding sales list, partial DU stock reduction, prescription screening, [FAR-06] sales returns, [FAR-07] sales returns per-register, [FAR-08] label, [SALES INFO], [FAR-09] prescription, [FAR-10] prescription details, [FAR-11] prescription-analysis per item, [FAR-12] sales (general), [FAR-13] sales details (general), [FAR-14] sales details (services), [FAR-15] general sales per R, [FAR-15] profit per doctor, [FAR-22] sales per distributor, [FAR-23] largest sales, [FAR-24] prescription screening, [FAR-16] sales returns (general), [FAR-17] sales returns details (general), [FAR-18] pharmacy revenue, [FAR-19] sales analysis, [FAR-20] recapped items.
3. Sales Menu Display: Registration number entry, personal data, drug type, instructions and the prescribing doctor, drawn from outpatient prescriptions by prescription receiving clerks to determine the prices. These data are useful for labelling, prescription screening and identifying the most frequently taken drug type.
4. Prescription Screening Menu Display: Clear prescription text, right patient, right drug, right dose, right route and preparation, right time and frequency, no duplication, no drug interaction, no allergy, no contraindication, PKU Muhammadiyah Hospital formulary suitability, BPJS suitability and insurance suitability.
5. Prescription Screening per Period Menu Display: Date, time, prescription code, prescription service, doctor, register, MR number, patient name, sex, patient address, age, cover, policy number, insurer, service, prescription screening, yes/no, note.
6. Prescription Screening Details per Period Menu Display: The prescription screening item would display overall outcomes with a recapitulation item taken from each date. This summary can be printed or imported into a spreadsheet as report material.
7. Prescription Screening Results Recap Display: Items of insurance suitability, BPJS suitability, PKU Muhammadiyah Hospital formulary suitability, right dose, right drug, right patient, right route and preparation, right time and frequency, no allergy, no duplication, no drug interaction, no contraindication, clear prescription text. These can be recapped with yes/no options which would yield a certain number in a particular period.

3.5. Prescription screening information system user perception

3.5.1. Descriptive analysis of user perception on the prescription screening information system usefulness

Outcomes of the descriptive analysis on the usefulness score data are presented in Table 5. Table 5 displays that statements 1 until 6, which denote items of user perception towards the utility of the prescription screening information system, yielded a total average score of 4.5 ±0.577 from 28 respondents. This implies that user perception in general over the usefulness of the prescription screening information system application was highly positive, in that the respondents benefited well from using the system.

A study found that excessive drug administration may lead to adverse drug reaction. Test implementation of a prescription screening information system revealed that respondents in this research were able to accomplish pharmaceutical care tasks to prevent adverse drug reaction [9]. Another investigation similarly indicated that respondents as users of the prescription screening information system considered it greatly beneficial in completing pharmaceutical care tasks at the hospital [10].

Table 5. Results of usefulness score data analysis (n=28)

No.	Statement	Average	SD
1	The prescription screening information system allows pharmacists to do their work at the hospital more easily.	4.571	0.504
2	The information system increases pharmacists' performance.	4.500	0.509
3	The prescription screening information system enhances productivity in pharmaceutical care at the hospital.	4.536	0.508
4	The prescription screening information system improves the effectiveness of pharmaceutical care at the hospital.	4.536	0.508
5	The information system makes pharmacists' tasks at the hospital easier to fulfil.	4.429	0.690
6	Overall, the prescription screening information system is useful for the completion of pharmaceutical care tasks at the hospital.	4.429	0.742
	Total average	4.500	0.577

3.5.2. Descriptive analysis of user perception on the prescription screening information system easy of use

Results of the descriptive analysis on the usability score data are given in Table 6. According to Table 6, statements 1 to 6, as items of user perception over the usability of the prescription screening information system, produced a total average value of 4.214±0.534 from 28 respondents. This implicates that the overall user perception towards the usability of the prescription screening information system application was largely favorable. It can thus be inferred that respondents found easy in using the system.

Table 6. Results of usability score data analysis (n=28)

No.	Statement	Average	SD
1	It was not hard for me to learn to use the prescription screening information system to support pharmaceutical care at the hospital.	4.250	0.441
2	I did not need much time to learn to use the prescription screening information system to support pharmaceutical care at the hospital.	4.179	0.548
3	Interaction with the prescription screening information system is easy to understand.	4.214	0.418
4	Interaction with the information system is flexible.	4.143	0.448
5	I can easily master the prescription screening information system.	4.250	0.701
6	Overall, the prescription screening information system is easy to use.	4.250	0.645
	Total average	4.214	0.534

In the previous research on electronic and non-electronic prescriptions showed the difference in skill between doctors who had long employed electronic prescriptions and those who had only recently. Results of a usability and aptitude test in the use of a prescription screening information system uncovered a respondent earning a score of 2 for lacking the ability to operate the system. Proficiency and ease in operating an information system are influenced by how long the user has interacted with it [11]. Another study reported that disclosed potential errors in prescribing and in drug interaction or allergy. The prescription screening information system could reduce potential errors in drug interaction and allergy as part of pharmaceutical care [12].

A study pointed out that the implementation of electronic prescriptions could control prescribing errors. It followed that the usage of a prescription screening information system to endorse pharmaceutical care could well decrease prescribing errors [13]. A different investigation demonstrated that the ease of use of information technology positively and significantly affected its usefulness. In practice, the application of the prescription screening information system was straightforward and greatly supported the realization of pharmaceutical care [14].

4. Conclusions

The main issue in manual prescription screening was its long duration. The adherence of pharmacists in conducting manual screening of 3,576 prescriptions in May, June and July 2015 was under 50%. The prescription screening information system design consisted of front display, pharmacy menu display, sales menu display, prescription screening menu display, prescription screening menu display per period, prescription screening details menu display per period and prescription screening results recap display. Perception over the usefulness of the prescription screening information system was largely positive.

References

[1] Gardner R M 2004 Computerized clinical decision–support in respiratory care. *Respiratory Care Journal* **49**(4) 378-388.

[2] Peraturan Menteri Kesehatan RI No. 012 2012 tentang Akreditasi Rumah Sakit. Retrieved from *www.depkes.go.id* on 27 April 2015.

[3] Soleymani F, Ahmadizar F, Meysamie A and Abdollahi M 2013 A survey on the factors influencing the pattern of medicine's use: Concerns on irrational use of drugs *J. Res. Pharm. Pract.* **2**(2) 59-63.

[4] Sujudi A 2004 *Keputusan Menteri Kesehatan Republik Indonesia Nomor 1027/MENKES/SK/IX/2004 tentang Standar Pelayanan Kefarmasian di Apotek,* Ministry of Health of the Republic of Indonesia, Jakarta.

[5] Fatta H A 2007 *Analisis dan Perancangan Sistem Informasi untuk Keunggulan Bersaing Perusahaan dan Organisasi Modern* (Yogyakarta: Penerbit Andi).

[6] Witten J L and Bentley C 2001 *System Analysis & Design Methods Second ed.* (Boston: Irwin Home Wood).

[7] Sujudi A 2004 *Keputusan Menteri Kesehatan Republik Indonesia Nomor 1027/MENKES/SK/IX/2004 tentang Standar Pelayanan Kefarmasian di Apotek,* Ministry of Health of the Republic of Indonesia, Jakarta.

[8] Jogiyanto 2005 *Sistem Teknologi Informasi* (Yogyakarta: Penerbit Andi) 32-37.

[9] Rambhade S, Chakarborty A, Shrivastava A, Patil U K and Rambhade A 2011 A survey on polypharmacy and use of inappropriate medication *Toxicology International* **19** 68-73.

[10] Budi S S 2016 Persepsi penggunaan terhadap kemanfaatan dan kemudahan penggunaan aplikasi sistem informasi baru *JBMA* **3**(1) 1-18.

[11] Shawahna R, Rahman N, Ahmad M, Debray M, Yliperttula M and Decleves X 2011 Electronic prescribing reduces prescribing error in public hospitals. *Journal of Clinical Nursing* **20** 3233-3245.

[12] Al Khajal K A J, Al Ansari T M, Damanhori A H H and Sequeri R P 2007 Evaluation of drug utilization and prescribing errors in infants: A Primary care prescription-based study *Health policy* **81** 350-357.

[13] Kaushal R and Bates D W 2002 Information Technology and medication safety: What is benefit? *Qual Saf Healt Care* **11** 261-265.

[14] Saputra E and Misfariyan 2014 Analisis penerimaan sistem informasi manajemen rumah sakit umum daerah bangkinang menggunakan *Technology Acceptance Model* (TAM) *Jurnal UIN Sunan Kalijaga.*

The effects of combination of *Eurycoma longifolia* Jack ethanolic extract and doxorubicine on hematological profile in rats given by 7,12-dimethylbenz(a)anthracene

L H Nurani[1], A Mursyidi[1], S Widyarini, A Rohman[3]

[1]Faculty of Pharmacy, Universitas Ahmad Dahlan, Yogyakarta, Indonesia

[2]Faculty of Veterinary Medicine, Universitas Gadjah Mada, Yogyakarta, Indonesia

[3]Faculty of Pharmacy, Universitas Gadjah Mada, Yogyakarta, Indonesia

E-mail: laelafarmasi@gmail.com

Abstract. Doxorubicin (Dox) is known as anticancer drug commonly used for cancer treatment. *Eurycoma longifolia* Jack or Pasakbumi was reported to have chemopreventive effect. In cancer patients, there are some dysfunctions of blood parameter, therefore some hematologic tests are needed to monitor cancer patients. In this study, the effects of combination of ethanolic extract of *E. longifolia* Jack (EEE) and Dox on hematologic profiles were investigated in rats injected by DMBA. Rats were divided into eight groups. Group I was normal group; Group II, rats were treated with extract dose 100 mg/kgbw; Groups III, IV, V, VI, VII and VIII, rats were treated with Dox, DMBA, DMBA+Dox, DMBA+EEE, DMBA+Dox +EEE, and Dox+EEE, respectively. DMBA administration orally was conducted twice a week for 5 weeks. At 16th week of treatments, bloods were taken from orbitalis sinus for hematologicals profile (levels of Hb, erytrocyte, hematocrite, leukocyte, MCV, MCH, and differencial leucocyte count) measurements. These data were analyzed by one way ANOVA followed by LSD test. DMBA administration significantly decreased the hematological profiles compared to the normal group, except in lymphocyte level. Rats treated with extract and extract+Dox were able to increase the hematological profile compared to rats given by DMBA only. Based on these findings it can be concluded that the combination of EEE and Dox potentially increase hematological profile of rats given by DMBA.

Keywords: 7,12-Dimethylbenz(a)anthracene, doxorubicin, *Eurycoma longifolia* Jack, hematological profile

1. Introduction

Eurycoma longifolia Jack, known as Pasak Bumi in Indonesia and Tongkat Ali, in Malaysia is belong to Genus of Eurycoma and Family of Simaroubaceae. *E. longifolia* Jack is one of the popular tropical herbal plants widely used for the treatment of diseases. It indigenous to South-East Asian countries like Indonesia, Malaysia, Thailand and Vietnam [1]. The dried *E. longifolia* Jack root is available commercially with price between 20 and 25 US dollars/kg, and the water extracts are sold in higher price, i.e. 26 US dollars per bottle of 60 capsules [2]. *E. longifolia Jack* is also formulated as herbal tea, known as "*E. longifolia Jack* tea"

Some important pharmacological activities have been reported for *E. longifolia* Jack, namely antitumor and anticancer, antimalarial, anti-diabetic, aphrodisiac, anxiolytic and anti-parasitic activities, as reviewed by previous studies [2, 3]. In addition, *E. longifolia* Jack has been used to treat a variety of angiogenesis related diseases such as cancer, rheumatoid arthritis, obesity, and psoriasis [4, 5]. These activities are mainly due to the active components present in *E. longifolia* Jack. These components are including of quassinoid, alkaloids, glycosides, eurycomanol, and eurycomanone [6, 7]. Quassinoids are anticancer agents exhibiting a significant cytotoxic activity [8] and increases spermatogenesis [9] .

Currently, the majority of health problems faced by human are cancer and the infectious diseases. The researchers and pharmaceutical industries have been trying to find out drugs either synthetic or herbal medicine to treat cancer [10]. In Indonesia, approximately of 8.2 million death is caused by cancers. It is estimated that the death due to cancer is 14–22 million for next 2 decades. Some dietary and medicinal plants, have been reported to possess substantial anticarcinogenic and antimutagenic effects to inhibit and treat the progression of cancer cells [11].

Some factors contribute to cancer, namely genetics, life styles, and carcinogens. One of chemical carcinogens belong to polycyclic aromatic hydrocarbon is 7,12-dimethylbenz(a)anthracene (DMBA) [12]. Doxorubicin (Dox) is known as anticancer drug commonly used for cancer treatment. Dox, however, revealed side effects toward normal cell. In order to reduce the side effects, doxorubicin is combined with plants extract, therefore the dose of Dox is reduced [13]. One of the potential plant used is *E. longifolia* Jack which has reported to have anticancer via anti-proliperative activity and apoptotic [14]. During chemotherapy, dysfunction of blood cell such as anemia [15] and thrombocytopenia [16] ussualy occurs. Therefore hematological profile test is needed. The aim of this study was to investigate the combination effects of *E. longifolia* Jack ethanolic extract and Dox on hematological profile.

2. Materials and Methods

2.1. Materials

The root of *E. longifolia* Jack was obtained from Martapura, South Kalimantan and the authentication was performed in Biology Laboratory, Universitas Ahmad Dahlan. The extraction procedure was conducted in the Laboratory of Pharmacy Biology, Universitas Ahmad Dahlan, Yogyakarta. The dried *E. longifolia* Jack was powdered and extracted using ethanol 70%. The 7,12-dimetilbenz(a)antracene (DMBA) was obtained from Sigma Aldrich (St. Louis, USA).

2.2. Animal treatment with DMBA, extract and doxorubicin

The animal treatment was performed according to previous study [17] using 56 female Sprague Dawley rats (60-70 g, 44 days old) which were purchased from Faculty of Pharmacy, Universitas Gadjah Mada, Yogyakarta, Indonesia.

Rats were randomly divided into 8 groups. Group I was normal group; Group II, which were treated with 100 mg/kg bw of EEE; Group III, which were treated with 1.12 mg/kg bw of Dox once a week for 5 weeks; Group IV, which were given by 20 mg/kg bw of DMBA orally, twice a week for 5 weeks. Groups V, VI, VII and VIII which were treated with DMBA + Dox, DMBA + ethanolic extract of *E. longifolia* Jack *(EEE)*, DMBA + Dox +EEE, and Dox +EEE, respectively.The rats were weighted once a week. At 16th week of treatments, bloods were taken from orbitalis sinus for hematological profile (levels of Hb, erythrocyte, hematocrite, leukocyte, MCV, MCH, and differencial leucocyte count) measurements

2.3. Hematological profile tests

Hematological profile tests were carried out in Laboratory of Parahita Diagnostic Center (a Clinical Laboratory), Yogyakarta, Indonesia by determining the levels of hemoglobin (Hb), erythrocyte (RBC), hematocrite (HCT), leukocyte (WBC), MCV, MCH, MCHC, RDW, thrombosis (PLT), lymphocyte, neutrophil, eosinophil and basophil.

2.4. Statistical analysis

Hematological profile of each groups were analysed using one way ANOVA using SPSS versi 16 for Windows, after being tested for normality of data distribution and homogeneity of variances. The significance level was set at 0.05.

3. Results and Discussion

Hematological profiles of rat given by DMBA and treated with *E. longifolia* Jack ethanolic extract, doxorubicin and combination of both are presented in the Table 1 and Table 2.

Table 1. Hematological profiles of rats given by DMBA and treated with *E. longifolia* Jack ethanolic extract, doxorubicin and combination of *E. longifolia* Jack ethanolic extract and doxorubicin

Group	Haemoglobin (g/dL)	Erytrocyte (million/μL)	Hematocrite (%)	Leukocyte (/μL)	MCV (fl)	MCH (pg)	MCHC (g/dL)
Normal	15.60±0.5	8.03±0.74	42.13±4.15	10162.50±3642.54	52.45±0.9	19.53±1.42	37.23±3.2
EEE	14.87±0.35	8.24±0.56	44.57±2.44	8796.67±1216.89	54.27±5.48	18.10±1.15	33.40±1.37
Dox	14.65±0.07	7.59±0.48	39.4±0.57	7045±1011.16	52±2.55	19.3±1.13	37.15±0.35
DMBA	14.70±1.29	7.40±1.09	41.88±4.11	8715.00±3902.35	57.27±6.91	20.05±2.97	34.95±1.36
DMBA+Dox	13.6±2.55	6.395±1.77	40.45±8.84	7120±254.56	63.75±3.89	21.55±2.05	33.75±1.06
DMBA+ EEE	15.33±1.19	7.80±0.73	42.60±1.37	7962.50±1219.93	54.83±3.51	19.70±0.66	35.95±1.67
DMBA+Dox +EEE	13.33±0.38	6.77±0.49	39.00±2.34	7836.67±793.87	57.67±1.55	19.80±1.42	34.23±1.7
Dox+EEE	7.75±1.06	3.985±0.76	22.45±3.89	3595±1152.58	55.7±0.14	19.3±0.71	34.6±1.27

EEE, ethanolic extraxt of *E. longifolia* Jack; Dox, doxorubicin; DMBA, dimethyle benz(a) anthracene; RDW, Red blood cell distribution width; RBC, red blood cells; MCV, mean corpuscular volume; MCH , mean corpuscular hemoglobin; MCHC, mean corpuscular hemoglobin

Table 2. Hematological profiles of rats given by DMBA and treated with *E. longifolia* Jack ethanolic extract, doxorubicin and combination of *E. longifolia* Jack ethanolic extract and doxorubicin

Group	RDW (%)	Trombocyte (/μL)	Limphocyte (%)	Neutrophyle (%)	Eosinofil (%)	Basophyle (%)
Normal	16.75±2.31	1159500.00±425555.72	60.50±7.59	31.50±7.77	1.67±0.58	0
EEE	17.83±1.04	846000.00±345813.53	63.33±2.08	17.33±13.43	6.33±3.79	0
Dox	17.7±0.28	1215000±427092.5	68±0	26±0	2±0	0
DMBA	16.48±2.42	1145833.33±269747.6	63.20±7.66	28.60±5.64	1.60±0.55	0
DMBA+Dox	23.3±5.80	1121500±204353.86	60±0	36±0	3±0	0
DMBA+ EEE	16.45±2.42	667750.00±175049.28	64.00±5.66	22.00±4.24	5.00±1.41	0
DMBA+Dox +EEE	15.37±2.89	517333.33±516946.16	49.50±14.85	21.00±9.9	2.00±0	0
Dox + EEE	13.85±1.48	403000±125865.01	60±1.41	17±8.49	4.5±4.95	0

EEE, ethanolic extraxt of *E. longifolia* Jack; Dox, doxorubicin; DMBA, dimethyle benz(a) anthracene; RDW, red blood cell distribution width; RBC, red blood cell

DMBA in the human body can cause oxidative metabolism of estrogenic hormones and result in some chemical metabolites such as reactive oxygen species (ROS) which capable to destroy the DNA. In addition, DMBA will attack hemoglobin to form haemoglobin adduct [18]. The hemoglobin adduct cause the level of haemoglobin reduce and, indeed, anaemia occurs. Anaemia can be identified by the reduced levels of MCV and MCH. Hemoglobin adduct is biomarker used to identify the presence of chemicals in the cell [19]. In the current study the rats given by DMBA revealed that the levels of hemoglobin, erythrocyte, hematocrite, leucocyte, MCHC, RDW, trombocyte, neutrophyle, eosinophile, basophil, MCV, MCH, and lymphocyte significantly lower than the normal group except for the leukocyte. The administration of EEE and Dox generally ameliorated the RDW, trombocyte, neutrophyle, eosinophyle, MCV, and MCH levels. In other hand, it decreased haemoglobin, erythrochytes, hematocrite, leucocyte, MCHC, and lymphocyte levels. These effects are attributed from the fact that haemoglobin causes oxidative reaction in haemoglobin. The *E. longifolia* Jack ethanolic extract can inhibit oxidative reaction via antioxidant mechanisms [20].

The erythrocyte levels of rats treated with EEE or Dox were not significantly different compared to DMBA group. However, the levels of erythrocyte in rats treated with EEE+ Dox reduced. Dox may reduce the interferon-γ (IFN-γ) production so that the erithropoesis and the formation of lymphocyte, monocyte, and neutrophil are inhibited [21]. The eosinophil levels in rats treated with DMBA were lower than those of normal groups, but it was not significantly difference ($p>0.05$).

4. Conclusion

The combination of *E. longifolia* Jack ethanolic extract and doxorubicin potentially increase hematological profile in rats given by DMBA.

Acknowledgement

The authors thank to the Ministry of Research and Higher Education of Republic Indonesia for financial support via competitive grant with contract number HB/III/2015.

References

[1] Khanam Z, Wen C S and HaqBhat I 2015 Phytochemical screening and antimicrobialactivity of root and stem extracts of wild *Eurycoma longifolia* Jack (Tongkat Ali) *J. of King Saud Univ.* Sci. **27** 23–30.

[2] Bhat R and Karim A A 2010 Tongkat Ali (*Eurycoma longifolia* Jack): A review on its ethnobotany and pharmacological importance *Fitoterapia* **81** 669–679.

[3] Rashid M, Kumar S and Ahmad B 2009 Medical Uses of *Eurycoma longifolia* Jack: a review *Pharmaceut. Res.* **2** 70–78.

[4] Kavitha N, Noordin R and Chan K L 2012 In vitro anti-toxoplasma gondii activity of root extract/fractions of *Eurycoma longifolia* Jack *BMC Complement. Altern. Med.* **12** (1–8) 91.

[5] Al-Salahi A, O S, Kit-Lam C, Abdul Majid A M S, Al-Suede F S R, Saghir S A M, Abdullah W Z, Ahamed M B K and Yusoff N M 2013 Anti-angiogenic quassinoid-rich fraction from *Eurycoma longifolia* modulates endothelial cell function *Microvascular Research* **90** 30–39.

[6] Bedir E, Abou-Gazar H, Ngwendson J N and Khan I A 2003 Eurycomaoside: a new quassinoid-type glycoside from the roots of *Eurycoma longifolia Chem. Pharm. Bull.* **51** 1301–1303.

[7] Low B S, Choi S B, Wahab H A, Das P K and Chan K L 2013 Eurycomanone, the major quassinoid in *Eurycoma longifolia* root extract increases spermatogenesis by inhibiting the activity of phosphodiesterase and aromatase in steroidogenesis *J. Ethnopharmacol.* **149** (1) 201-7.

[8] Jiwajinda S, Santisopasri V, Murakami A, Kawanaka M, Kawanaka H, Monique G M, Eilas R, Balansard G and Ohigashi H 2002 In vitro anti-tumor promoting and anti-parasitic activities of the quassinoids from *Eurycoma longifolia*, a medicinal plant in Southeast Asia *J. Ethnopharmacol.* **82** 55–58.

[9] Low B S, Das P K and Chan K L 2013 Standardized quassinoid-rich *Eurycoma longifolia* extract improved spermatogenesis and fertility in male rats via the hypothalamic-pituary-gonadal axis *J. Ethnopharmacol.* **145** (3) 706-14.

[10] Ali Ahmed H E, Abdel-Salam H A and Shaker M A 2016 Synthesis, characterization, molecular modeling, and potential antimicrobial and anticancer activities of novel 2-aminoisoindoline-1, 3-dione derivatives *Bioorganic Chemistry* **66** 1–11.

[11] Youssef K M, Ezzo A M, El-Sayed M I, Hazzaa A A, EL-Medany A H and Arafa M 2016 Chemopreventive effects of curcuminanalogs in DMH-Induced colon cancer in albino rats model *Future J. Pharmaceut. Sci.* **1** 57-72.

[12] Adventus B, Tiny E H 2013 Ekstrak metanol daun kelor menurunkan ekspresi BCL-2, TRAIL-R1, dan kadar caspase-3 jaringan kolon tikus yang diinduksi DMBA *J. Kedokteran Brawijaya* **27** 4.

[13] Ang H H, Hitot S Y, Fukaya H and Takeya K 2002 Quassinoids from *Eurycoma longifolia Phytocemistry* **59** 833-837.

[14] Mahfudh N and Azimahtol HLP 2008 Eurycomanone induces apoptosis through the up regulation of p53 in human cervical carcinoma cells *J. Cancer Mol.* **4** (4) 109-115.

[15] Dicato M, Plawny L and Diederrich M 2010 Anemia in cancer Annals of Oncology **21** (Supplement 7) vii167–vii172.

[16] Kuter D J 2015 Managing thrombocytopenia associated with cancer chemotherapy. *Oncology J.* **29** (4) 282-94.

[17] Jenie R I, and Meiyanto E 2007 Ko-kemoterapi ekstrak etanolik daun sambung nyawa (*Gynura procumbens* (Lour.) Merr.) dan Doxorubicin pada sel kanker payudara *Majalah Farmasi Indonesia* **18** (2) 81-87.

[18] Ogawa M, Oyamate, Isse T, Yamaquchi T, Murakami T, Endo Y and Kawamoto T 2006 Hemoglobin adducts as a marker of exposure to chemical substances, especially PRTR class I designated chemical substances *J. Occup. Health* **48** (5) 314-28.

[19] N'jai A U, Larsen M, Shia L, Jefcoate C R and Czuprynski C J 2010 Bone marrow lymphoid and myeloid progenitor cells are suppressed in 7,12-dimethylbenz(a)anthracene (DMBA) treated mice *Toxicology* 505341-505349.

[20] Rifkind, Joseph M, Joy G, Mohanty, and Enika N 2015 The pathophysiology of extracelluler hemoglobin associated with enhanced oxidative reactions *Front. Physiol.* **5** 500.

[21] Zhang, Xiao-Yu, Li, Wen-Guang, Wu, Yong-Jie, and Gao and Ming-Tang 2005 Amelioration of doxorubicin-induced myocardial oxidative stress and immuosuppression by grape seed proanthocyanidins in tumour bearing mice *J. Pharm. Pharmacol.* **57** (8) 10431.

Relationship of therapeutic outcome with quality of life on type 2 diabetes mellitus patients in Abdul Azis Singkawang hospital

D A Perwitasari[1], S Urbayatun[2], I N Faridah[1], N Masyithah[1]

[1]Faculty of Pharmacy, Universitas Ahmad Dahlan, Yogyakarta, Indonesia

[2]Faculty of Psychology, Universitas Ahmad Dahlan, Yogyakarta, Indonesia

E-mail: diahperwitasari2003@yahoo.com

Abstract: Diabetes is one of the diseases that required long treatment. Therapeutic outcome is one of the important factors that affect the quality of life. The purpose of this research is to know the effect of therapeutic result on quality of life in Abdul Azis Singkawang hospital. This study used Cross-sectional design. The inclusion criteria for this study was patients with type 2 diabetes mellitus (T2DM) outpatients over 18 years with ICD code X E.11. This study used the EQ-5D to measure patient's quality of life. We recruited 86 T2DM patients who met the inclusion criteria and were dominated by female respondents around 57%. The average value of quality of life EQ-5D was the index value 0.75±0.22 and visual analog scale 74.02±11.80. The result of the analysis showed that there was significant relationship between income and quality of life (p=0.001) and there was significant correlation between 2-hour PG and quality of life (p=0.037). The conclusion of this study was the therapeutic outcome affect the quality of life in 2-h PG, where the higher 2-h PG showed the low quality of life.

Keywords: Quality of life, EQ-5D, therapeutic outcome

1. Introduction

Diabetes mellitus (DM) is one of the most common endocrine diseases that become a health problem in the worldwide. American Diabetes Association states that DM was a group of metabolic diseases with characteristics of hyperglycemia due to insulin secretion abnormalities, insulin action and both [1]. Type 2 Diabetes mellitus was one of the most common types found in more than 90-95% [2]. Indonesia is estimated in 2030 will have a DM of 21.3 million people [3].

Quality of life is the individual's perception in relationtship to the purpose of life, hope and attention. DM is a chronic diseases that required therapy continuously so can affect the quality of life patient's and can lead to tendency DM patients had been complications [4]. Quality of life described various scopes such as a physical health, psychological conditions, degree of independence, social relationships, personal beliefs and aspects related to situations felt, according to their individual developmental environment [5].

The most common method that used to measure was the EQ-5D questionnaire (European quality of life-5 dimensions). The EQ-5D questionnaire consisted of 5 dimensions of mobility, self-care, regular activity, pain/discomfort, and anxiety/depression, plus a global question to assess the general

IOP Conf. Series: Materials Science and Engineering **259** (2017) 012023 doi:10.1088/1757-899X/259/1/012023

health condition called VAS (visual analog scale) [6]. The desired of therapeutic outcome was to change the lifestyle/diet (weight control and physical activity) and using drugs antidiabetic or insulin [7]. If not handled properly, complications due to diabetes can occur was microvascular complication (retinopathy, nephropathy and neuropathy, mainly on sensory nerves such as gangren, autonomic nerves such as sexual dysfunction and gastroparesis), macrovascular complications (CHD, cerebrovascular diseases, PAD) and others complications (psychosocial problems, dental diseases) [7]. The purpose of this study is to determine the effect of therapeutic outcome to the quality of life.

2. Materials and Methods

This study used cross-sectional design and prospective data retrieval in diabetic patients at Abdul Azis Singkawang hospital.

2.1 Tools and materials

The data was collected used the EQ-5D (European quality of life-5 dimensions) questionnaire. Patient characteristics include age, sex, education, income, occupation, length of treatment, long suffering, family history of DM, social status, comorbidities and therapies used were collected. Treatment outcomes include levels of BG, FBG and 2-h PG measured using the Easy Touch tool, and HbA1c measurements using DCA Vantage Analyze instruments. Inclusion criteria include type 2 diabetes mellitus patients, aged >18 years, had complications, and outpatients, while the exclusion criteria include patient refusal, illiterate patients and type 1 diabetes mellitus.

2.2 The way of research

The target population in this study were all patients with T2DM (ICD X E.11) which were outpatients and treated at Abdul Azis Singkawang hospital from January to March 2017. This study used primary data from EQ-5D questionnaire.

2.3 Statistical analysis

In this study the statistical analysis used univariate analysis which describe the characteristic of each variable, bivariate analysis knew the relationship between two variables, multivariate analysis knew the relationship of some variables. The reliability test in this research used Cronbach's method alpha coefficient and validity test in this research used content validity, Pearson correlation. Content validity was fulfilled if correlation coefficient correlation above 0.30 [8].

The EQ-5D questionnaire consisted of 5 questions and 1 global question to assess the state of public health called VAS (visual analog scale). The 3 level EQ-5D assessment started from 11111 to 33333 and the index value ranges from 0-1. A value of 0 indicated that health was worse than death, whereas if value 1 indicated good quality of life [9].

3. Results and discussion

3.1 Demographic characteristics

Characteristics of the patient include sex, age, status, education, occupation, income, diseases duration, type of treatment and comorbidities. The distribution of T2DM patients by characteristics can be seen in Table 1.

Based on Table 1, it can be seen that most of T2DM patients were female (57%). This was in line with previous research results where T2DM patients with female sex were higher when compared with men [14]. The mean age of T2DM patients was 58.16±7.76 years, this was because the aging process was caused decrease ability of β-pancreatic cells in producing insulin [10]. The highest percentage of education level was higher education (64%). The average distribution of income was 2.5±8.75 million, meaning the average income of patients was above the minimum sallary rate of Singkawang (2.2 million). Around 90.7% of patients received oral therapy with treatment duration of 6.70±4.90 years, and diseases duration from 7.31±5.38 years. Most patients in this research had 57% macrovascular diseases complications of heart, gangrene, hypertension and stroke.

Table 1. Patients characteristic

Characteristic (n=86)	X±SD	
Age (years)	58.16±7.769	
Treatment duration (years)	6.70±4.90	
Diseases duration (years)	7.31±5.38	
HbA1c (%)	9.50±2.62	
FBG (mg/dl)	158.69±70.49	
BG (mg/dl)	244.74±95.19	
2-h PG (mg/dl)	244.99±89.63	
Characteristic (n= 86)	**Amount**	**Percentage (%)**
Gender		
a. Male	37	(43%)
b. Female	49	(57%)
Education		
a. Low	31	(36%)
b. High	55	(64%)
Occupation		
a. Jobless	33	(38.4%)
b. Occupied	53	(61.6%)
Income		
a. <2.2 million	30	(34.9%)
b. >2.2 million	56	(65.1%)
Marietal status		
a. Married	76	(87.4%)
b. Single	10	(11.6%)
Family history		
a. Yes	53	(61.6%)
b. No	33	(38.4%)
Type of treatment		
a. Insulin	8	(9.30%)
b. Oral hypoglycaemic	78	(90.7%)
Comorbidities		
a. Microvascular (kidney, eye)	12	(14.0%)
b. Macrovascular (heart, gangrene, hypertension, stroke)	49	(57.0%)
c. Macrovascular & Microvascular	25	(29.1%)

Note. FBG = Fasting blood glucose; BG = Blood glucose; 2-h PG = 2 hours prandial glucose; HbA1c = Hemoglobin A1c.

3.2. Associations between sociodemographic and quality of life

The results of univariate analysis which showed the mean value of quality of life level in patients with type 2 diabetes mellitus can be seen in Table 2.

Table 2. Patient quality of life rate

Dimention of EQ-5D	Mean	SD
Index	0.7591	0.22969
VAS	74.0233	11.80426

Note. VAS = visual analog scale; Index = the state of health patient's

The result of univariate test analysis showed that the quality of life in patients with T2DM in Abdul Azis hospital was good, it can be seen from the index valued close to 1, while the VAS also showed a good quality of life.

The relationship between characteristic and quality of life can be seen in Table 3. Based on bivariate result there was no significant relationship ($p<0.05$) between sex with quality of life, meaning that gender does not affect the quality of life. There was no significant difference ($p<0.05$) between jobless and occupied of quality of life. There was no significant difference ($p<0.05$) between have family history and no family history. There was no significant difference ($p<0.05$) between type of treatment of insulin and oral hypoglycaemic, meaning that theraphy did not have relationship with quality of life. Some studies explained that patients with T2DM who consumed insulin induced diseases burden and reduced satisfaction with treatment of patients used orally treatments [11].

Table 3. The relationship between characteristic and quality of life

Characteristics	n	Index		VAS	
		Mean ± SD	*p*	Mean ± SD	*p*
Gender					
Male	37	0.77±0.17	1.000	75.6 ± 12.00	0.297
Female	49	0.74±0.26		72.7 ± 11.60	
Education					
Low education	31	0.71 ± 0.25	0.138	69.83 ± 9.80	0.005*
Higher education	55	0.78 ± 0.21		76.39 ± 12.27	
Occupation					
Jobless	33	0.73 ± 0.25	0.577	74.70 ± 13.41	0.431
Occupied	53	0.77 ± 0.21		73.70 ± 10.81	
Income					
<2.2 million	30	0.68 ± 0.25	0.015*	69.17 ± 11.22	0.005*
>2.2 million	56	0.76 ± 0.21		76.62 ± 11.36	
Family history					
Yes	53	0.79 ± 0.19	0.110	74.16 ± 10.36	0.871
No	33	0.70 ± 0.26		73.78 ± 13.97	
Marietal status					
Married	76	0.76 ± 0.22	0.665	73.69 ± 11.81	0.631
Single	10	0.73 ± 0.30		76.50 ± 12.03	
Type of treatment					
Insulin	8	0.69 ± 0.35	0.765	65.62 ± 16.78	0.097
Oral hypoglycaemic	78	0.76 ± 0.21		74.88 ± 10.96	
Comorbidites					
Microvascular	12	0.79 ± 0.12	0.536	72.58 ± 6.94	0.062
Microvascular	49	0.76 ± 0.24		76.42 ± 12.07	
Micro+ Macrovascular	25	0.73 ± 0.24		70.00 ± 12.24	

Note. VAS = visual analog scale; *$p<0.05$

Quality of life was strongly inflouced by the severity of complications of T2DM diseases [11], but in patients in RSUD Abdul Azis found no significant difference between comorbidities and quality of life. There was significant difference ($p<0.05$) between low education and high education. Patients who are highly educated are better qualified than those with low education [15]. This can be interpreted as a positive effect of education for that matter T2DM management.

There was significant difference ($p<0.05$) between <2.2 million and >2.2 million, that low quality of life was associated with the low socio economics of T2DM patients [16].

3.3. Associations between sociodemographic and therapeutic outcome

From Table 4, the level of education (p=0.000) influence one of the outcomes of T2DM patient therapy was FBG, this was because patients with higher education level of knowledge and understanding was greater, so they better understand how to controlled blood glucose well. There was significant (p<0.05) between income and therapeutic outcome, this showed that income has an influence on FBG (p=0.002). There was significant (p<0.05) between family history and therapeutic outcome. Family history have an effect on the patient's 2-h PG (p=0.047).

Other factors that can cause the increase of blood glucosa levels include hormones, genetic disorders and eating patterns [17]. There was significant difference between type of treatment and HbA1c indicated with p value 0.048 and the significant difference between type of treatment and BG with p< 0.001. This suggests that the type of treatment received by an outpatient T2DM patient does not affect the value of blood glucose. Marietal status divided into 2 was single and married which can be seen that it has relationship with outcome of therapy in patient T2DM. There was significant differences between 2-h PG (p=0.019) and FBG (p=0.031) in T2DM patients with marietal status. This suggests that marietal status conditions was related to the outcome of therapy, but it was also found that diabetes mellitus patients are chronic diseases that required significant behavioral changes, supported by high family attention, and low family conflicts can be a high level of adherence [12].

Table 4. Mann Whitney test results analysis between characteristics with therapeutic outcome

Characteristic	HbA1c		FBG		BG		2-h PG	
	X ± SD	p	X ± SD	p	X ± SD	p	X ± SD	p
Education (n)								
a. Low education (31)	9.91±2.66	0.280	194.45± 77.9	0.000*	258.39±108.2	0.416	266.9±95.40	0.150
b. High education (55)	9.34±2.68		138.53±57.4		237.05±87.11		232.64±84.60	
Income (n)								
a. <2.2 million (30)	10.21±2.74	0.089	189.13±78.20	0.002*	257.43±115.15	0.584	267.93±89.87	0.077
b. >2.2 million (56)	9.19±2.59		142.38±60.64		237.95±82.93		232.70±87.84	
Family history (n)								
a. Yes (53)	9.87±2.53	0.117	165.28±76.89	0.520	253.47±93.87	0.297	258.21±89.60	0.047*
b. No (33)	9.02±2.85		148.09±58.33		230.73±97.05		223.76±86.83	
Marietal status (n)								
a. Married (53)	9.76±2.72	0.053	164.46±72.27	0.031*	249.50±95.42	0.153	252.26±89.89	0.019*
b. Single (33)	7.94±1.57		114.80±31.21		208.60±89.76		189.70±68.53	
Type of treatment (n)								
a. Insulin (53)	11.25±2.29	0.048*	249.88±85.73	0.001*	276.00 ±110.54	0.308	310.50±122.12	0.067
b. Oral hypoglycaemic (33)	9.37±2.66		149.33±62.16		241.54±93.70		238.27±83.79	

Note. HbA1c = Hemoglobin A1c; FBG = fasting blood glucose; BG = blood glucose; 2-h PG = 2 hours prandial glucose; *p<0.05

3.4 The relationship between therapeutic outcome,characteristic and quality of life

The result of multivariate relationships between quality of life, characteristic and therapeutic outcome are showed in Table 5.

Table 5. Results of regression linear therapeutic outcome and characteristics of quality of life

Variable dependent	Variable independent	Coeffisient of determination (r^2)	Corellation (r)	p
Indeks	Income	0.051	0.225	0.037
VAS	Income	0.087	0.253	0.001
	2-h PG	0.139	-0.265	

Note. VAS = visual analog scale; 2-h PG = 2 hours prandial glucose

Based on Table 5, income significantly influence quality of life was index of 5.1%, income and 2-h PG significantly give effect to quality of life was VAS of 8.7%. The results of linear regression

IOP Conf. Series: Materials Science and Engineering **259** (2017) 012023 doi:10.1088/1757-899X/259/1/012023

analysis showed that there was relationship between the outcome of therapy with T2DM patient quality of life. The result of negative correlation on income was index and VAS (r=0.225; r=0.253) means that the higher the income, the higher quality of life. The socio-economic status associated with DM affects a person to perform his self-care management [18] so that quality of life was also affected. VAS was influenced by the 2-h PG indicated by the negative correlation (r=-0.265) value means higher 2-h PG, quality of life scores was lower. Blood glucose level was one of the factors that affect the quality of life [19]. The results was aligned in previous studies that the better the control of blood glucose then the quality of life of T2DM patients the better [20]. Besides the control of blood glucose was very important to be done to assess the quality of life of T2DM patients because most affect the patient's functional, psychological and social health [13], with the increasing rate of lifestyle diseases such as diabetes mellitus in many developing countries, healthcare policy makers should develop policy to educated diabetes patients to have a good glycemic control.

Limitations of this study are the limited number of respondents who match the inclusion criteria and the need of time to explained the intent or content of the questionnaire.

4. Conclusion

We found that 2-h PG and income could affect quality of life. The result of negative correlation analysis showed that higher 2-h PG showed lower VAS value, while the higher the patient's T2DM income, so the higher quality of life.

Acknowledgement

The author would like to thank the head and hospital staff who assisted in this research. The authors was also grateful to the Ministry of Technology and Higher Education for initial funding and sponsoring this research, also thank to all the patients who participated in this research.

References

[1] ADA 2010 Diagnosis and Classification of DM Mellitus *DM Care* **33** 562-569.
[2] Center for Disease Control and prevention (CDC) 2014 *Diabetes report card, national center for chronic disease prevention and health prevention*. 3-4.
[3] Kementerian Kesehatan RI 2013 *Riset Kesehatan Dasar* (Jakarta: Badan Penelitian dan Pengembangan).
[4] Rizkifani S, Perwitasari D A and Supadmi W 2014 Pengukuran kualitas hidup pasien DM di RS PKU Muhammadiyah *Farmasains* **2**(3) 2.
[5] Romero M, Consuelo D V and Guzman N A 2013 Is Health Related Quality of Life (HRQoL) a Valid Indicator for Health Systems Evaluation? *Springer Plus* **2** (664): 1-7.
[6] Zeng Y C, Shirley S Y C and Alice Y L 2010 Quality of life measurement in women with cervical cancer: implications for chinese cervical cancer survivors *Health and quality of life outcomes* **8** 30.
[7] ADA 2011 Diagnosis and classification of diabetes mellitus *Diabetes Care* **34** 62-9.
[8] Nugroho I S 2010 Hubungan pembelajaran pendidikan kewarganegaraan dengan kesadaran hukum siswa x sma negeri kartasura tahun ajaran 2008/2009 *Skripsi* Fakultas Keguruan dan Ilmu Pendidikan Universitas Sebelas Maret Surakarta.
[9] Agborsangaya C B, Markus L and Jeffret A J 2014 Comparing the EQ-5D 3L and 5L: measurement properties and association with chronic conditions and multimorbidity in the general population *Health and Quality of Life Outcomes* **12** 74.
[10] Fitriyani 2012 Faktor risiko dm mellitus tipe 2 di Puskesmas Kecamatan Citangkil Dan Puskesmas Kecamatan Pulo Merak Kotacilegon *Skripsi* Universitas Indonesia.
[11] Romulus T, Lulian Ve, Bogdan T, Diana L, Christian O, Deiana R and Octavian M 2016 Factors influencing the quality of life perception in patients with type 2 DM mellitus *Dovepress* **10** 2471–2477.

[12] Abdel W A, Jude U O, Shafika A A A and Adel M T 2006 DM patients' family caregivers' subjective quality of life *Journal of National Medical Association* **98**(5) 1-10.
[13] Zurdayanis and Erline M 2010 Hubungan kadar glukosa darah dengan kualitas hidup pada pasien diabetes melitus tipe II di RSUD Sleman Yogyakarta *JKKI* **2**(6) 2-3.
[14] Kayyis H 2016 Penilaian persepsi terhadap penyakit dan kualitas hidup pada pasien diabetes melitus tipe 2 dengan komplikasi kronis di RSUD Pringsewu Lampung *Tesis* Universitas Ahmad Dahlan Yogyakarta.
[15] Tchicaya A, Nathalie L, Stefan D, Jean B and Danieal R W 2015 Relationship between self-reported weight change, educational status, and health-related quality of life in patients with DM in Luxembourg *Health and Quality of Life Outcomes* **13** 149.
[16] Gautam Y, Sharma A K, Agarwal A K, Bhatnagar M K and Trehan R R 2009 A cross sectional study of QOL of diabetic patient at tertiary care hospital in Delhi. *Indian Journal of Community Medicine* **34**(4) 346-350.
[17] Sustrani L, Alam S and Hadibroto L 2004 *DM* (Jakarta: PT. Gramedia Pustaka Utama).
[18] Puspitasari P 2014 Gambaran tingkat pengetahuan dan sikap tentang monitoring kadar gula darah mandiri pada pasien DM di RS PKU Muhammadiyah Yogyakarta *Skripsi* Program studi ilmu keperawatan fakultas kedokteran dan ilmu keseharan Universitas Muhammadiyah Yogyakarta.
[19] Nissa M K 2013 Hubungan kadar glukosa darah dengan kualitas hidup penderita diabetes melitus tipe 2 di Rumah Sakit Umum Daerah (RSUD) Kota Cilegon periode Januari-Mei 2013 *Skripsi* UIN Syarif Hidayatullah Jakarta.
[20] Khanna A, Bush A L, Swint J M, Peskin M F, Street R L and Naik A D 2012 Hemoglobin A1c improvements and better diabetesspecific quality of life among participants completing diabetes self management programs: a nested cohort study *Health and Quality of Life Outcomes* **10**(48) 2-6.